WATER BYWAYS

WATER BYWAYS

DAVID E. OWEN

WITH PLATES AND MAPS

DAVID & CHARLES : NEWTON ABBOT

0 7153 5981 9

Set in 11 on 13pt Times Roman
and printed in Great Britain
by Latimer Trend & Company Ltd Plymouth
for David & Charles (Holdings) Limited
South Devon House Newton Abbot Devon

LIST OF ILLUSTRATIONS

Plates

*Photographs not acknowledged above are from the author's
collection*

Maps

INTRODUCTION

I HAVE written enough to show my love of the quiet waterways and the peaceful countryside through which they flow. We find the relaxation of a boating holiday on the canals is a perfect antidote to a busy life. We love the early mornings when we share the fields with sheep and cattle and the solitude of the evenings when the birds come in to roost. Yet these canals which contribute to our peace of mind were nearly lost to us before we could enjoy them. Built for commerce during the Industrial Revolution, their trade went first to the railways and then to the roads. Many were actually abandoned, some filled in and others left to moulder away.

In 1946 Robert Aickman and a few others formed the Inland Waterways Association to save and restore the waterways network for all purposes. They saw that water carriage could relieve the pressure on the roads, particularly in towns, that water was a vital commodity and that there was room for all to enjoy the canals. Had it not been for them, many more canals would have been abandoned and lost for ever.

Volunteer organisations will always be needed to watch over the canals. While the 1968 Transport Act preserved for enjoyment a considerable part of the total network, it took away the age-old right to navigate. When the ten year fight to restore the Cheshire Ring of canals appeared at last to be won, a breach occurred in the Bridgewater canal so disastrous that there was even talk of abandonment. When our relations with the British Waterways Board who controlled the canals had become most friendly and co-operative, the Government issued plans to split the canal system and hand over to independent Water Boards. Nevertheless, the future of canals seems brighter now than ever before. This is because so many people care desperately to restore, preserve and guard them. When the Inland Waterways Association was first formed, it seemed to be a voice crying in the wilderness. Those who wanted to save canals were even referred to by a Minister of Transport as the 'lunatic fringe'. Now it is admitted by all that countless boaters, anglers, ramblers and nature lovers turn to the canals to enjoy unspoilt countryside. There is even a Minister for the Environment.

To my mind the most exciting development in the canal and river scene is the emergence of the volunteer conservationists. First, Mr C. D. Barwell collected helpers and funds and bought and restored the beautiful lower Avon river from Tewkesbury to Evesham. Then David Hutchings rebuilt the Stratford-upon-Avon canal using working parties of volunteers, prisoners from Winson Green gaol, sappers and others from the Armed Services and finished a monumental task which was opened to the public by Her Majesty Queen Elizabeth the Queen Mother in July 1964. This was a demonstration to all that volunteers could have the skills and the ability to rebuild locks and refit paddles and gates. Volunteer groups began to spring up in many parts of the country ready and able to clean up and restore lengths of canal. At this point, Graham Palmer started *Navvies Notebook* which recorded the names and addresses of working party organisers, their programmes and site locations and gave details of the canals on which they were working. Furthermore, it made possible the mobilisation of large bodies of volunteers to tackle particular tasks such as Operation Ashton, the canal clearance at Welshpool and the Dudley Dig. Working parties were soon visiting other areas to help in specific work and all the while their experience and technical abilities were improving. This was soon acknowledged by the British Waterways Board who have co-operated closely.

At the time of writing, the greatest piece of restoration by volunteers is the rebuilding of the Upper Avon navigation again under the charge of David Hutchings. This fell into disuse one hundred years ago and is being completely rebuilt with new locks in different places, new weirs and newly dredged channels. Furthermore it is being carried out by a Trust whose funds have been voluntarily subscribed. In this, *Navvies Notebook* (shortened to *Navvies* and now the publication of the Waterways Recovery Group) has been able to organise the volunteers and keep a continuous programme of work.

What of the future? In the 1960s we were working to save, clear and restore canals which were threatened with closure. These include the Caldon, that beautiful little navigation which wends its way into the quietest valleys of the Pennines; the Ashton and Lower Peak Forest which form so vital a link in the Cheshire ring of canals; the Erewash in Derbyshire; the huge Kennet and Avon which connects the Thames to the Bristol Channel and many others too numerous to mention.

Now, with the lead given by David Hutchings on the Upper Avon and the organisation provided by Graham Palmer and *Navvies*, many people are thinking seriously of reviving navigations long since lost. The Welshpool canal abandoned in 1944 which flows through the superb country of the Welsh Borderland, close to the Severn Valley and beneath the Breidden Hills could be reopened; the Grantham canal from Nottingham through the Vale of Belvoir, closed in 1936, would be a splendid addition to the waterways network; and thoughts are turning to others up and down the country which seemed so recently to be beyond repair.

I believe we shall need all these rivers and canals in future as our expanding industrial community turns to the countryside for fresh air and recreation. I also believe that our own personal involvement in their restoration is producing something which will become very dear to us. In the world today, it is easy to destroy or turn away with indifference as others litter and pollute. The rebuilding of navigations with our own hands in our spare time makes us realise that they form a heritage to be cared for and handed on to our children.

This book tells of further tours on canals and rivers of England in our 40ft steel boat *Rose of Sharon*. It describes also some of the most exciting examples of volunteer restoration. The crew includes my wife, Pearl, and our little beagle, Lindy, still very active particularly around the locks despite her eleven years. We welcome Jennifer and Roger, our daughter and son who join us occasionally for a few days at a time. It is usually just the two of us, however, who have not yet been turned back without reaching our goal—and our goals have extended beyond the better-known routes into some of the quieter canals which can truly be called 'water byways'.

DAVID E. OWEN

CHAPTER ONE

THE LEEDS & LIVERPOOL CANAL

WE had known the Leeds & Liverpool canal for some years, had worked a friend up the Wigan locks in 1963 and had been to Liverpool, Blackburn, Burnley and Chorley with the National Rally of boats in 1965. We had seen the famous Bingley five-rise and walked the towpath at Skipton. We had cruised the beautiful Marton Pool with one friend and had run past the Aintree racecourse with another. But we had never taken *Rose of Sharon* on the canal and this we determined to do. At Whitsun we took her from her home mooring on the Macclesfield canal down the Cheshire locks through Middlewich and Preston Brook and along the whole length of the Bridgewater canal. We turned north at Waters Meeting, crossed the Irwell on the Barton swing aqueduct and moored at Worsley. There we left her for three weeks, incidentally attending the weekend Worsley rally. We planned to rejoin her in mid-June starting our holiday on the Thursday evening.

We had everything packed ready, had eaten a quick tea when I came home from work and were waiting for a taxi to collect us in a violent thunder storm. We had all sorts of stores to take for the journey together with hand baggage and Lindy. We were held up for half an hour as the taxi was lost trying to find us and we finally set off about six thirty. By the time we reached Worsley the rain had stopped and we were able to go aboard and cast off in the evening sunshine. Our intended destination for the night was the length of Leeds & Liverpool canal between Leigh and Wigan often referred to as the 'Moon Country'.

The Leigh branches of both the Bridgewater and Leeds & Liverpool canals were built on coal measures and have suffered considerable subsidence and in consequence the channel is extremely deep throughout. We made good progress past collieries which were still using the canal for transport and on through the town of Leigh.

Leigh surprised me for I had expected it to be a coal town which it certainly was but not a cotton town. Its dominant features from a distance are the great chimneys from mills which were built on the canalside. It seemed a pity that they did not use canal transport for their raw materials and for their finished products for barges carrying fifty tons can reach the docks of either Liverpool or Manchester. Perhaps they neither spin nor weave today but surely some of their wares could still travel by water in such a deep and well-maintained navigation. We cruised on through a small bridge past the inconspicuous junction into the Leeds & Liverpool canal.

I had been told to look out for Plank Lane bridge, a heavy swing bridge worked automatically by a bridgekeeper. I thought it was on the outskirts of Wigan and it was with some surprise that we found it just beyond Leigh. I had not realised we were there and went ashore to push it myself. 'You can't move that, luv' came a voice from the cottage below. The bridgekeeper was working in his garden and he came up and shifted it for us holding up a bus as he let us through. We moored for the night in surprisingly pretty country on an embankment built up as the canal had subsided, overlooking a flash with reeds and swans. Like so many canal moorings, it was a remote spot and we had it all to ourselves.

I described the history of the Leeds & Liverpool canal in *Water Rallies*, how it was first planned in 1766 and finally opened throughout its whole length fifty years later. Brindley made a survey and Longbotham was the engineer when the Act was obtained in 1770. For me it was of especial interest as an

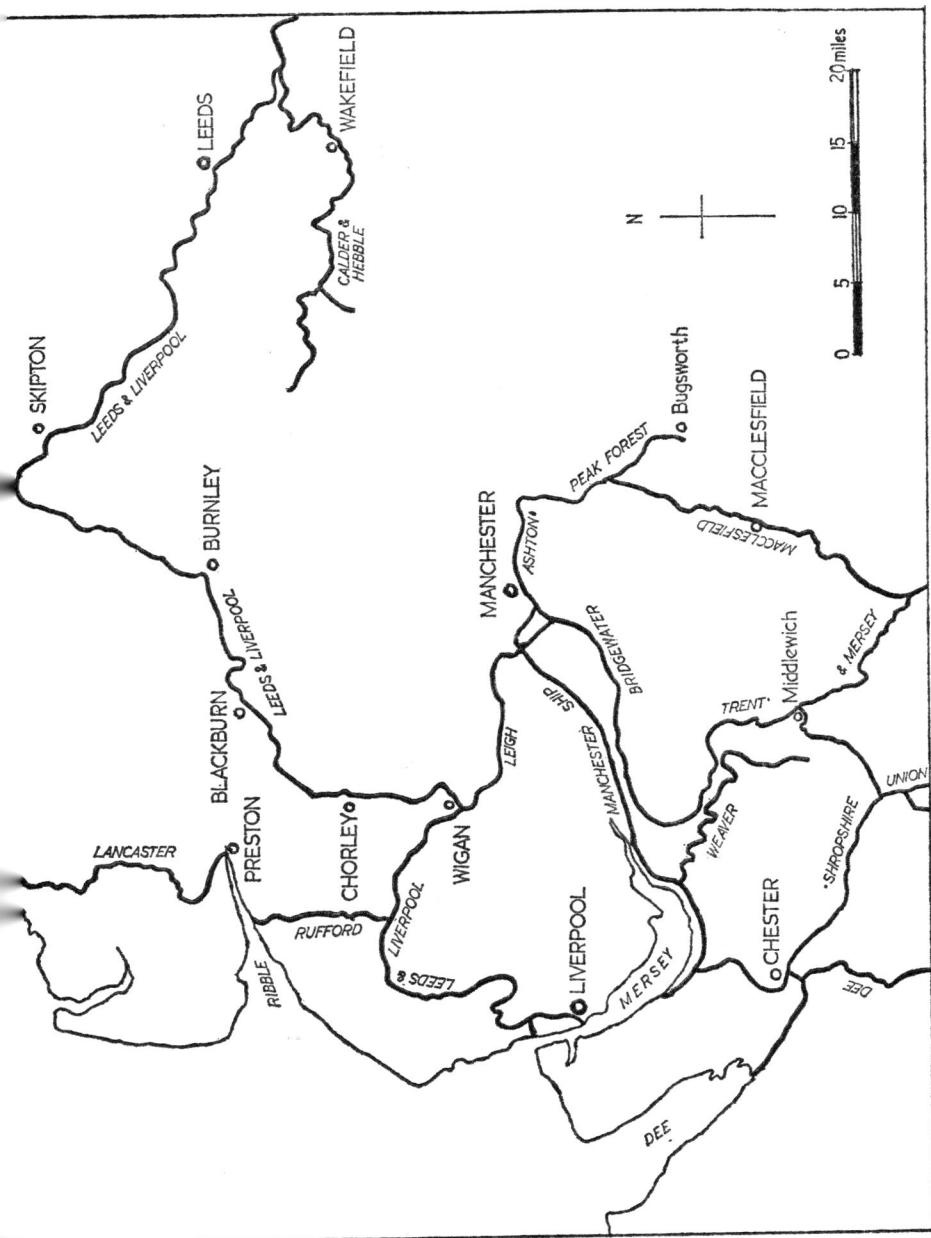

Northern canals and navigable rivers

example of the other method of canal building. Phillips in *A General History of Inland Navigation,* published in 1792, described Brindley's surveys for the 'grand cross', the canals which were to link the ports on the Trent and Humber with those on the Mersey, with branches to the Severn and the Thames. These canals were constructed on the narrow gauge of 7ft lock width to conserve water needed in the heavy lockage as they passed over and through the hilly country of the Midlands. They were and still are used by narrowboats which are quite unsuitable for work on the tidal estuaries and at sea. Because of this, inland ports like Shardlow and Stourport grew up for trans-shipment of goods to coastal craft.

At the time of this survey there was an alternative plan which chose a much easier route. It was to start at Northwich on the already navigable Weaver and cross Cheshire via Winsford, Minshull, Madeley Park and Whitmoor to Stafford. Thence it was to join and follow the Trent valley to Burton and to enter the river which was then navigable to that point. It was to have a branch via Eccleshall and Wellington to the Severn near its junction with the river Tern. The locks were to be 60ft long and 13ft wide and no tunnels were needed. The advantages of this route were the cheaper construction and locks which suited the coastal and estuary craft of the day, the Mersey Flats, the Humber Keels and the Severn Trows. This would mean that no trans-shipment was necessary but there were two great disadvantages. The line of these canals did not pass near the rapidly developing Midland industrial towns—Stoke-on-Trent, Wolverhampton, Birmingham and Coventry. Thus they would receive no backing from industrialists who were most in need of canal transport. Further, they terminated in river navigations and in the case of both the Severn and the Trent the upper parts of these navigations were not very successful and are not, in fact, navigable today. Brindley's plan won the day and much of the country was committed to the narrow lock and the narrowboat.

That Brindley was right at the time seems to be confirmed by the Leeds & Liverpool canal though we may well regret the narrow gauge today. This was planned with short, wide locks, 14ft 6in wide by 62ft long, quite suitable at the time for Keels and Flats, and the original course from Leeds, also surveyed by Brindley, followed the Aire valley westwards to Skipton and Gargrave, crossed into the Ribble valley at Whalley, skirted the hills to Parbold and continued across the Lancashire Plain to Liverpool. A branch was to run up to Wigan. The total length was 109 miles and the estimated cost just over a quarter of a million pounds. It left literally high and dry the developing Lancashire cotton towns.

Longbotham pressed ahead with the work and the eastern end was open through Skipton and Gargrave in 1777. The western end was easier, for use was made of the river Douglas navigation from Burscough as far as Wigan. The original terminus in Wigan had the expressive name of 'Mirey Lane End' and this was connected to Liverpool by canal in 1775. At this stage the money was used up and nothing further was done until 1790 when Whitworth was called in to advise. It was at this point that the mistake in cutting out the cotton towns was acknowledged, for the raising of further money depended largely upon them. The route was lengthened by some 18 miles, the great canal tunnel of Foulridge was planned and the line took in Nelson, Colne, Burnley, Blackburn and Chorley. It was then to run along the hillside and drop down a great flight of thirty locks to Wigan.

Even now the work proceeded very slowly reaching first Burnley, then Blackburn from the east. In 1792 an Act was obtained to build a canal from Kendal through Lancaster to Preston and across the Ribble estuary to a second summit at Walton. Thence it was to go through Chorley to the outskirts of Wigan. This was planned by Rennie and the two sections were built on either side of the Ribble to be linked, temporarily it was hoped, by a light railway. The southern section from Walton would run parallel to the Leeds & Liverpool canal at a lower level. It was obviously ridiculous to build the two canals side by side and terms were soon agreed with the Lancaster Canal Company for the Leeds & Liverpool canal to drop through

B

seven locks at Johnsons Hillock and use the Lancaster canal to Aspull above Wigan, dropping down from there by twenty-three locks to their western length at Mirey Lane. The whole 127 miles was finally opened throughout in 1816. Five years later the Leigh branch was built to link up with the Bridgewater canal and from there the rest of the country. In 1822 the locks on the Leigh branch and the length from Wigan through to Parbold were lengthened to take full-length narrowboats. At Appley, below Wigan, the older, shorter locks are still visible and just usable. In 1846 a connection was made with the Mersey at Stanley Dock.

We looked forward to exploring this canal and its branches and early next morning, for we had a big day ahead of us, we continued towards Wigan. We passed 'Dover lock', now removed because submergence had made it unnecessary, and reached the lower of the two locks on the outskirts of Wigan at about ten o'clock. It was a sunny morning, already very warm, and we had the twenty-three wide locks ahead of us. The bottom lock was open and to our delight we found a British Waterways official standing by to help us. It was notable on the whole length of the Leeds & Liverpool canal that the British Waterways staff of foremen, lock-keepers and others helped us whenever it was consistent with their other duties. Further, we received a most friendly welcome from the engineers at their Wigan offices when we called in on our way.

Most of the paddle gear on the Leeds & Liverpool canal is secured against vandalism by chains and padlocks and a screw key must be obtained from the Wigan offices en route. Furthermore there are a few stiff and heavy paddles which call for a lock windlass with a rather longer handle and this, too, can be obtained. In any case it is well worth notifying the Wigan office in advance for the lock-keepers can then be told and will give as much help as possible.

Beyond the two locks on the Leigh branch we reached the junction with the main canal and faced the great flight of twenty-one locks which lift the canal out of the Mersey valley on to the high ground of mid-Lancashire. Here our daughter, Jennifer, joined us. She is a great enthusiast and loves working locks, so had taken a day off specially to tackle them and to help us up. In her case it was her third passage having taken her own boat through to the Blackburn Rally in 1965 and helped a friend's boat up two years earlier. Actually, thanks to the Waterways staff, our task was made easy and Jennifer was little more than a very welcome passenger.

For a student of canal engineering the paddle gears on the Wigan locks deserve mention. This is also true of many of the other locks along the length of the canal. In most cases both ground and gate paddles are at the top of the lock and many of the former are opened by a huge lever which must be lifted from the ground to a vertical position. These are known as 'cloughs' (pronounced to rhyme with 'cows') and with the weight of water against them they can be very heavy. Furthermore, the whole structure must be dropped before the gate is open fully. The commonest alternative is a sluice opened by worm gear with a fixed handle over the top. When fully open the handle is secured with a chain to stop it flying back. When the paddle is dropped the handle is released and it is then advisable to stand well back. Many of the gate paddles are drawn back by a handle and toothed wheel on a long ratchet and once more the weight of water may make them stick. Some of the paddles are being replaced by modern types which are lighter to work. However, on the whole length of the canal we never found paddles too heavy to move or in fact as stiff as some we had met on other parts of the canal system—notably above Leicester and on the river Stort.

I have described on page 171 the easiest way to pass through the locks on the Leeds & Liverpool canal. On the Wigan locks we placed the boat against the towpath side and raised only the paddles on the same side as ourselves, the ground paddle first. We came up the flight in three and three-quarter hours. Perhaps we could have saved a little of this time by raising all the paddles at once, but we should certainly not have had so easy a passage.

Throughout the journey the sun rose higher and the day became hotter and long before we had reached the top we had worked up a grand thirst. There are three pubs near to the top locks and our friends from British Waterways had their own preference. It was already past our lunch time and we sat round in a cool bar parlour eating sandwiches while they told us tales of the canal and bygone days. The canal boatman is a natural teller of tales and I only wished we had had a tape recorder to hold and retell the tales in our own home.

Once at the top of the Wigan flight the scenery changes completely from coal mines and pit heaps to open country. We were soon passing through parkland which lies in front of Haigh Hall with its rhododendron bushes ablaze with colour. After 4 miles we reached Adlington where Jennifer left us.

Next morning we cruised through Chorley, recalling the rally of boats five years previously. Now the scene had completely changed for Botany Bay, where we had embarked the mayor, was now restricted by the new motorway, and the towering railway viaduct which had crossed high over the moorings had vanished. I remembered seeing pictures of the demolition of this structure, the piers and arches disintegrating as the explosive charges were detonated. Beyond Chorley we reached the bottom of Johnsons Hillock locks.

We moored and walked along the remaining quarter mile of the old Walton Summit as far as the first bridge beyond which the canal was filled in. The motorway, the main cause of the abandonment, rose on an embankment beyond us and cars flashed by at a pace so different from that of the quiet canal. We wished that there had been the money to build the aqueduct across the Ribble in John Rennie's day so that we could have reached the quiet waters of the northern length of the Lancaster canal. This water link was never completed and the Lancaster canal has remained isolated except for those who have trailed their boats or braved the sea passage to Glasson Dock. Now even the 3 mile Walton Summit has almost completely gone.

The Johnsons Hillock flight of seven locks rises through

delightful country and once more we came up gently using only the paddles on the towpath side. At the top are moorings for boats and a fine lock-keeper's house with a water-point, chemical toilet disposal and all 'mod. con'. We were pleased to find the Leeds & Liverpool canal well furnished with these amenities. In addition the lock-keeper was as friendly and helpful as all others we met on the canal. The village of Wheelton was near at hand with excellent shops.

Our next town was Blackburn and we cruised on to the Ewood embankment with more memories of the 1965 rally. We thought again of the long line of decorated boats and the mayor walking along the towpath amongst the crowds in his scarlet robes. We reached the locks and worked slowly up through them.

It was Sunday morning and we had no milk. Pearl reckons that I have a nose for milkmen and a persuasive tongue for the lady in the newspaper shop who might be able to spare a pint. I walked across to the main road but there were no shops open and nothing in sight. All at once I saw a milk van rounding a corner and moving away in the opposite direction so I chased after it and caught it. Once more I was able to return to *Rose of Sharon* triumphant with fresh milk which tastes so much nicer than any substitute in tea.

The canal above Blackburn locks is urban and unattractive for some 2 miles. More recently I have heard that the Blackburn Rotary Club is putting into effect plans for its improvement. Rounding a bend near the cathedral we saw a miserable dog covered with slimy weed trying desperately to reach the towpath from the water. It was nearly exhausted and had such a pathetic look in its eyes. We made for the bank and I grabbed it just in time and pulled it out. The water along the towpath is fairly deep and we saw for the first time some ingenious structures for extricating horses which had fallen in. The towpath was cut back and a double cobbled slope ran down into the canal. We were told that if a horse fell in, it would swim to the slope where it could be led out. The slopes were in both directions and it did not matter which way the horse was going.

It is about nine miles from Blackburn to Burnley by road and seventeen by water, the canal winding round the sides of both hills and valleys. At Church Kirk we could see a canal bridge across the valley a few hundred yards away but had to cruise nearly two miles to reach it. On this pound we met the first of many swing bridges which form such a feature of the Leeds & Liverpool canal.

We moored for the night on the slopes of Hameldon Hill from whose quarries came the stone for the bridges and canal banks. Away to the north were the slopes of Pendle but we were not disturbed by the witches who are said to reside there.

Before entering Burnley next morning we cruised through deep woodlands and Pearl called my attention to a kingfisher darting ahead of us. It landed on a tree stump and allowed us to cruise right up to and past it without moving. It sat with its shoulders hunched in the position in which they are always shown in illustrations.

Burnley proved a handy place for shopping and for fuel but we were soon away along the famous mile-long embankment which carries the canal through the centre of the town well above rooftop height. This is one of the seven wonders of the waterways and we were to pass through three others on this trip—the Bingley five-rise, the Barton swing aqueduct and, digressing on our way home, the Anderton boat lift. It would have needed too great a diversion to have included the Pontcysyllte aqueduct near Llangollen on this trip and the other two, the Standedge tunnel and the Devizes flight of locks are not at present navigable to *Rose of Sharon*.

Beyond Burnley are Nelson and Colne, but we were soon to reach the Barrowford locks which carry the canal into the superb Craven highlands. One little bridgehole appeared to be blocked by a small tug but it moved out of our way as we approached. As we passed, however, it appeared to be aground and the steerer commented sadly, 'I'm on t'bottom gettin' out of yer rowad.'

The Barrowford locks are beautifully cared for and again the lock cottage at the top of the flight has all the amenities. We worked up behind a hire boat and though the lock-keeper emptied the locks ahead of us, we were not prepared for the rush of water from the overflow of the lock below the top. This was coming down with terrific force making it quite impossible for Pearl to steer for the lock chamber. The lock-keeper hurried to the top to find that the hire boat crew were trying to fill the lock with the bottom paddles open! This was soon put right and all went well.

Half a mile beyond Barrowford locks is the Foulridge tunnel, wide and fairly high and little short of a mile in length. It is here that the legendary cow is said to have fallen in and swum the entire length to be pulled out the other end. After revival with a bottle of brandy, it is said to have been none the worse for the experience! Beyond the tunnel is Barnoldswick-in-Craven where monks from Fountains Abbey first settled before moving on to Kirkstall near Leeds. They were there for five years and complained of the inhospitable country and the terrible weather. The summer rains were so continuous that the corn sprouted on the stalk and their cattle were stolen by wild Scots who swept in from the north. The weather treated us rather better for we had a fortnight of unbroken sunshine and there was no sign of the marauding Scots.

This 5 mile length is the top pound of the canal and is little short of 500ft above sea level. On waterways at present navigable, this elevation is exceeded only on the Macclesfield and Peak Forest, the Caldon and the short Titford length of the B.C.N. From here the canal drops steadily down into the Aire valley and follows the river to join it at Leeds. At the end of the pound are the three Greenberfield locks standing out in the rich Craven countryside. Mr Wilkinson, the lock-keeper, invited us in to see his charming house and garden, below which runs the old course of the canal and this is said to have dropped a three-rise staircase to the next pound. Engineer Longbotham who constructed this length cannot have considered the waste of water and it was not long before it was replaced by the present trio of locks and only the line of the old channel and a single bridge remain.

Below Greenberfield locks stretches the 5 mile Marton pool, perhaps the prettiest length on this or any canal in the country. The canal winds backwards and forwards round the sharpest corners so acute that posts were erected with vertical rollers for the towropes, for the horse was round the corner long before the boat had reached it. This is Mountain Limestone country with green fields and white stone walls. Half-way along the pound is East Marton bridge, an extraordinary structure with one arch standing on top of another and up the hill is a fine hostelry famous for its welcome as well as its food and drink. We very stupidly timed our passage wrongly and were not able to test it out for ourselves.

At the end of the pound the canal drops gently through seven Bank Newton locks past the village of Gargrave and on to the 17 mile Skipton pound. Until that point, we had been almost on our own but we shared these locks as far as the wharf at Gargrave with three other boats, two of which were making for the Trent. Between two of the locks the canal crossed over the river Aire on the fine stone Priestholme aqueduct. We moored and looked over at the river, still a mountain stream rippling over boulders and rock ledges. We then explored Gargrave, a lovely stone village, typical of the Yorkshire Dales.

As we passed a farm above the bottom lock, a young man waved and followed us. He was Philip Green, son of the farmer, and a mine of information on the whole canal. He was a most useful contact and we soon became firm friends.

Our next town was Skipton, ancient and gracious with fine shops and a noble castle. The mooring is in the less attractive part a little distance from the wide main street and the shopping centre. From the mooring and running northwards parallel to the main street, however, is the little Springs branch constructed to carry limestone from the quarries which lie behind the castle. We had noted it from the map and Philip had told us that whatever we did we must not miss it. He warned us that it might be weedy and that there was probably not room for us to turn our boat at the terminus, but we are used to these hazards and cruised in.

The branch was built under an Act of Parliament of 1773 and now runs through some of the prettiest parts of the town. It passes under a stone bridge and curves round beneath the towering castle walls. The terminus is a small basin shaped like a quarry, with chutes for loading the limestone into the boats. As we cruised along, the duckweed became thicker until at the end it looked firm enough almost to bear our weights. At the extreme end we tried to turn but there was not sufficient width and we decided to retreat backwards. For a few hundred yards this meant poling for the duckweed built up against our stern, but soon we cleared it and came out using the engine. It had certainly been worth the trip for it took us into the loveliest parts of the old town.

Skipton is on a level pound of 17 miles and across this are nineteen swing bridges. Most are fairly small, built only for farm tracks, but a few carry important roads. I was able to swing all but one without help, but this carried a main road 2 miles east of Skipton and needed two of us to move it. On our return, a group of children were waiting to swing it and we did not need to go ashore to help. We soon worked out a drill with Pearl bringing the boat near to the towpath so that I could go ahead to the bridge. If the bridge was busy, I would have to wait for a lull in the traffic and then swing it across. The heaviest bridges had buffers to receive them when they were fully open to the canal and I had to be careful not to bounce the bridge too hard or it would start to swing back. I had to take especial care when I had enthusiastic local helpers and heard of one boatman whose bridge bounced back so far that the boat only just avoided a collision.

We moored for the night a short distance beyond Skipton against the towpath. I stress 'against the towpath' for across the canal was a masked bull—one of the most fearsome sights I have ever seen. We sincerely hoped that he could not swim.

At the end of the Skipton pound is the Bingley five-rise, the great staircase of five locks which drops the canal 60ft in only 80yd. This fine structure was something we had come particularly to see. I had read all the instructions and knew that if I found the locks full, as they would have been if someone had just come up, I should have to go to the bottom and empty the bottom lock. I should then have to go to the second lock and empty it into the bottom lock and re-empty the bottom lock. I would then go to the third lock and empty it into the second lock and so on. . . . Eventually all would be empty except the top lock which would be ready for me to enter. This is all necessary because you cannot empty a full lock into another full lock without flooding the flight. At Foxton, there are side ponds and in the staircase of two at Etruria on the Caldon canal, there is an overflow. Here, there must have been a terrific waste of water when the canal was heavily used with boats going in both directions.

I carry a rather super windlass with ball-bearings in the handle and I took this to prospect. By the lock cottage were the lock-keeper and his friend and they looked me up and down. In the manner of Al Read, the lock-keeper said, 'Thinkin' of goin' down?' I said 'yes'. 'Then you won't need that for a start.' He was right, of course, for all the paddles have fixed handles. It then turned out that he knew we were coming and had prepared all the locks for us. Like all we met on the Leeds & Liverpool, he was very helpful indeed.

Coming up a couple of days later, the section inspector gave us a hand. I came ashore and Pearl steered the boat in and threw the ropes up to me. It is a long throw for the stern rope but the lock wall rises towards the lock above and the bow rope has to be flung much higher. After making one or two unsuccessful attempts to catch it, we found it easier for me to keep it ashore with me. There is a trick of passing the rope beneath the footbridges which I have often attempted with varying success. A sufficient length is hung down on the far side of the bridge and is then swung right over in a great circle so that the end can pass under the bridge and up on to it. It is easy in theory and simple when done by an expert, but when the boat is moving and there is time for one attempt only, the rope is likely to catch on its way over or come up in the wrong place. A helpful boatman remarked to me that *this* was the way to do it. I was foolish enough to reply that I already knew it. 'Oh, you know it, do you,' he said, 'then catch it!'

We had help down the Bingley five-rise but quite enjoyed doing the Bingley three-rise by ourselves. There are several more staircases of two and three locks before Leeds is reached and we realised how wasteful of water they were. Engineer Longbotham must have thought that the water supply was limitless in the Craven Pennines, a mistake which was to worry the company many times in the long history of the canal.

We decided to turn soon after we had passed Shipley for we wanted time to explore the other end of the canal and we knew the eastern end of the navigation well from our Leeds days. I used to look after Kirkstall Abbey, a fine Cistercian ruin on the banks of the Aire within sight of the canal. We spent a night in the woodlands above Fieldhead locks before returning to Shipley. The day before we turned was the occasion of the General Election and we felt we must stay awake long enough to hear some of the results. The woodlands were so peaceful and we were so tired and full of fresh air that we were asleep by ten o'clock and knew nothing until the seven o'clock news next morning.

Shipley proved a good shopping centre and we looked at the end of the derelict little canal which once stepped up into Bradford. On our way back along the Pennine lengths of the beautiful waterway we received the same friendliness and help from the lock-keepers and waterways staff and were pleased to use the well-organised amenities. We were sorry to leave it and drop down through Wigan to the western lowland length which runs through to Liverpool.

We spent our last night on the top near to the parkland of Haigh Hall and a young couple came down from a cottage to chat to us on the towpath. When we arrived at the top lock next morning, the lock-keeper was waiting. Some people on a hire-boat had started even earlier than we and had already passed the first lock. 'I catched them in the top lock,' said the lock-keeper, who held them up so that we could lock down together and save water. The boats fitted snugly side by side and we had a very easy descent.

Below Wigan the canal runs through countryside scarred by coal-mining. The little river Douglas first made navigable in 1720 lies a short distance to the south. The original termination Mirey Lane End occurred to us several times as we looked over the scene. We dropped through Pagefield and Ell Meadow locks, the first of the short locks which had been replaced by ones long enough to accommodate full-length narrowboats and cruised on out into more attractive country.

The Douglas valley and the Leeds & Liverpool canal run through a gap in the hills with the beautiful wooded Parbold Beacon on one side and the bolder Ashurst Beacon on the other. Before debouching on to the plain they pass beneath the motor-way M6 and we recalled the many times driving north when we had seen a flash of water in the valley far below. Almost directly beneath the road are the two Dean locks side by side. As we approached, we came on a number of people sitting round the lock and found that they belonged to an art group or class sketching the lock from all sides. Our appearance was hailed with delight for now they could add a boat to their scene.

The last lock before the 27 mile level pound which winds across the rich lowlands of Lancashire is Appley Bridge. At this point a single modern lock of great depth and with steel gates replaces two older locks of shorter length whose channel lies alongside.

Our next port of call was Parbold out in the plain with its beautiful bridge and ancient windmill so often pictured in canal literature. The village is well worth a visit and we were able to replenish all our stores from the well-stocked shops. Beyond Parbold we crossed the river Douglas on a long embankment with a small aqueduct over the river. Moored for the night on our way back we found that we could see Blackpool tower for, though the canal is less than 100ft above sea level, the views are wide across the flat lands.

Burscough comes next, dominated by Ainscough's huge flour mill. In the days when the canal was more thriving commercially, Ainscough's had their own fleet of boats carrying to and from the docks of Liverpool.

At Scarisbrick, the canal makes its nearest approach to the sea with buses running to Southport only 4 miles away. Near Ormskirk, the canal runs in a small cutting and it is said that here the first sod was cut in 1770. At Haskayne we visited the famous Ship Inn where Mr and Mrs Pip Dunn welcome all canal-lovers and run a fleet of hire cruisers. At Lydiate we moored by the Mersey Motor Boat Club headquarters, the narrowboat site in the National Rally of boats in 1965.

I described in *Water Rallies* the final stretch of the Leeds & Liverpool canal from Lydiate through Maghull past the Aintree Race Course and through Bootle to the locks which lead down to Stanley dock in Liverpool. Since the National Rally, the Transport Act of 1968 has designated 8 miles of this length a 'Remainder Waterway'—a canal with an uncertain future. The IWA held a full-scale boat rally in Liverpool the same year to show how vital it is to retain and develop this length and the Merseyside section has produced a pamphlet describing how replanning could improve the amenities of the whole area.

The canal is well constructed and in excellent condition and obliteration and filling in would not solve the problem for it carries water to many industries and drains a wide area of surface waters. Alternative water supplies and additional deep sewers would cost millions of pounds to construct. Furthermore the canal connects directly with the Liverpool docks and could carry a large tonnage of goods to a base near the M6 motorway, thus relieving the city of much heavy traffic. If the canal is retained, such a development must surely come.

Against the retention of the canal are many whose chief concern is the safety of children and it is true that most children love water. A canal that is little used, difficult of access and cut off from public routes by high walls, is always a potential danger. There are no adults about and there have even been cases of children drowning in sight of people who were unable to reach them. Records show that a busy canal with boats on the water and people on the towpath is very much safer than a forgotten waterway. The Leeds & Liverpool canal would be very much safer along its final 8 miles if it was opened out and fully used. I remember visiting Crosby where the town clerk told me that every school child in the borough has the opportunity to learn to swim in the town's swimming pool. This is surely the greatest insurance against drowning children. Further, there is no proof that the children would be safer if the canal was filled in and replaced by a road, for roads take a terrible toll of young lives. It is easy to exclaim after a drowning, 'The canal is a menace, fill it in.' Apart from the cost and other problems, this is seldom the right way to save life and limb.

On our route back to Wigan we determined to explore the Rufford branch which follows the river Douglas and enters the tidal reaches at Tarleton. We asked a number of people what they knew of its condition. We ought to have known better or, at least, we should have chosen the right people to ask. 'Don't go down there, there's a boat stuck in the middle and you'll never get through,' 'It's too narrow to turn, you'll have to come out backwards,' 'The weed is something terrible and there's no depth at all,' they said. Even the IWA makes it an adventure canal, no doubt to attract people to it. However, at least we were sensible enough not to be put off. We phoned Jennifer who said she would like to join us and collected her on Sunday morning at Burscough.

We turned in at Burscough at the beautiful wide bridge which bears the date 1816. For the first mile the wide but not very deep locks lie close together. The lower ones are spaced more widely and they lead off to Rufford. There is a wide section below Rufford lock with deep water right up to the bank which we chose for a night mooring on our return journey. This gave access to the main road and we were able to arrange for Jennifer to be collected at the end of the day. Below this is a long level stretch to the sea lock at Tarleton. We passed the gardens of Rufford Old Hall and had a little difficulty with swing bridges which grated on gravel as they were seldom moved. The sides of the canal were shallow and I had to use my long shaft to help me to vault ashore.

The last 2 miles make very lovely cruising for the canal takes on the appearance of a Broadlands river winding through the flat rich countryside. Far away behind us were the summits of Parbold and Ashurst Beacons and we had a huge expanse of sky with fleecy white clouds. Some stretches were covered with duckweed but this had no effect on our air-cooled engine.

We finally reached Tarleton and moored above the sea lock. As always on these occasions the tide was out and we could not have locked down into the tidal stretches even if we had wanted to do so. Our only problem in turning was a boom placed across the canal to stop the advance of duckweed. We ran our bows on to it and had to pole back carefully.

We winded the boat and made for our morning at Rufford, a village which has a happy memory for me when I went shopping. Pearl asked me to buy some eggs and I brought half a dozen, all of which turned out to be double-yoked! I wished I had bought more.

The next day there were only the two of us to carry on up the locks. There was a terrific cross wind and I had to go ashore in plenty of time to swing the bridges so that Pearl could keep weigh on and hold the boat in the channel. With a particularly stiff bridge which I did not open in time, I had to take a rope to pull the bows into the gap. I had also to prepare the locks and open them in good time for Pearl to cruise in under full power. In one lock we found the British Waterways men waiting for us with the gates open. As we had cruised down on the Sunday, they had no idea we were on their length and they were delighted to see us.

Eventually we reached the junction at Burscough and turned

into the main canal. We were pleased to have seen the little Rufford arm which did not bear out any of the odd statements we had heard. It was an adventure canal but alas! no adventures except those caused by a stiff cross wind which can make steering difficult anywhere. This is a little navigation worth exploring for its own sake and essential to retain as a link with the sea.

The next day we cruised back through Parbold and into Wigan, turning right to make for Leigh and the Bridgwater. As we came through the lock before the turn a young man came across to join a boat ahead of us. With heavy rain starting we wanted to share the two locks on the Leigh branch so we made for the junction at full speed. As we started to turn we met a fully laden coal barge taking his corner widely to get the maximum depth of channel. There looked to be no room for us to pass between him and the towpath and it was too late to swing away in front of him. He waved us on and we squeezed through scraping sides but our steel hull saved us from any damage. We met two more laden barges in the locks and finally moored well out in 'moonland' in torrential rain.

The memory we shall retain of the Leeds & Liverpool is of a beautiful wide deep canal, well maintained and with every amenity; a canal of clear clean water running by green hillsides; a canal of great engineering with splendid flights of locks, fine embankments and aqueducts; and especially a canal with friendly North Country people, of helpful lock-keepers and waterways staff. When we were cruising it was very much underused. I believe a time will soon come when many more people will explore its reaches and discover its charm.

Page 33   Leeds & Liverpool Canal: (*above*) entrance to Rufford Arm;
(*below*) Kildwick showing one of the many swing bridges

Page 34   Dudley Canal: (*above*) the Delph locks showing the old line on the left; (*below*) Castle Mill Basin in the Dudley tunnel

CHAPTER TWO

THE STOURBRIDGE AND DUDLEY CANALS

WE had cruised the Birmingham Main Line from Aldersley to Worcester Bar and had found it an easy but rather unattractive route. It climbs the Wolverhampton twenty-one locks on to the 473ft level and continues by way of the Coseley tunnel and the Tiptons lock to the long, straight cuttings of Birmingham. We were taking *Rose of Sharon* from Norbury where she had been dry docked and were making for the Grand Union. It seemed a good chance to plan our route to take in the Stourbridge and Dudley canals. We set off at Whitsun in the hot sunshine with the heavy scent of May blossom wafting from the hedgerows.

There had been some doubt as to whether we should be able to follow this route at all as one of the locks in the Stourbridge flight had been damaged a fortnight before by an over-wide converted narrowboat. The British Waterways Board staff had been prompt in making a complete repair and we were assured of a passage. We were amused at other completely groundless warnings—'You'll need a long-handled windlass for those paddles,' 'You want a good strong dog to protect you up that way,' and 'Oh! you'll never get your boat up there.' They were reminiscent of an old retired Cheshire boat woman's warning on Stoke, 'It's a wicked place, that's where they drown the young girls when they've done with them.'

We reached Autherley, turned southwards down the Staffs &

Worcs canal and started to drop down the locks. Once past the impressive Bratch, not quite so brightly painted as when we had last seen it, and the small, deep Botterham staircase of two, lined with detergent foam, we ran through the woodlands bordering the river Stour. We had forgotten that this length is so beautiful and were pleased to be cruising it again. With the little cuttings in the deep red rocks and the sunshine filtering through the intense greenery above, it was colourful and beautiful. We passed the end of Ashwood basin, now a marina, but used for over one hundred years for coal traffic from the important collieries including Shutt End on Pensnett Chase. A railway had been built to bring the coal down and the last coal boats finished about twenty years ago. When we had passed in 1961, it had looked unused and forgotten, but now there were signs of activity and a small boat cruised out ahead of us.

It was lunchtime on Whit Sunday when we moored at Stourton Junction. We had come down through eighteen locks in order to climb a further twenty-nine to reach the Birmingham level but the detour was well worth while. We remained moored for an hour after lunch during a brief but sharp thunderstorm and at three o'clock set off up the four Stourton locks on to an attractive, wooded pound. It was still sunshine and showers when we reached the bottom of the Stourbridge sixteen locks where a number of children stood hopefully awaiting our arrival. On the right below the bottom lock is a small sandstone cliff and a branch of the canal runs off to the right to Stourbridge. Into this we turned to the surprise of the children who shouted that we could not go up there. We recalled the Stourbridge rally of boats in 1962 when David Hutchings had brought his equipment from the Stratford canal and had dredged the branch to the summit.

The first 200yd were narrow and shallow but we made steady progress. A narrowboat was moored by some pretty cottages and this restricted the channel but we were just able to edge past. A few feet beyond this we were firmly aground and through the crystal clear water we could see that we were not going to get much farther. We tried poling for some minutes but could move neither forward nor backward and then I caught the hook of my long shaft in some hidden obstruction and could not pull it out. I heaved so hard that eventually the boat began to move and the hook was finally dislodged. We continued to pole backwards for a while and then reversed out of the arm having been approximately 300yd up and back in an hour and a half! We cruised back a quarter of a mile towards Stourton for a peaceful night's mooring.

Next morning dawned bright and sunny and we were up early. I told Pearl that all we needed to do was to turn the boat in what appeared to be a slightly wider section and we could be at the bottom of the Stourbridge locks in no time. Pearl suggested reversing the quarter of a mile but I would not hear of it. We cast off, reached the place I had marked, and turned the bows into the reeds. Carefully poling the stern we were soon jammed across the canal and it was obvious the boat was about a couple of feet too long. Pearl again suggested reversing but I still scoffed at the idea. 'We'll find a spot in no time,' I said, 'and it will be much quicker than moving slowly back in reverse.'

Half a mile farther on we found another possible turning place, but once more jammed across the canal. When we were most tightly wedged, a boat appeared which had earlier come down the sixteen locks. We held it up before we were able to get clear again and it was infuriating to think of sixteen empty locks awaiting us and we were still the wrong way round! Pearl forbore to comment as we cruised on seeking a winding hole and half expecting to meet a boat which would take the empty locks.

I was beginning to wonder if we should be able to turn before we reached the Stourton locks or if we should have to drop down four locks in order to face the other way! At last we reached a wide enough place and eased the boat round. Once more we held up the boat which had appeared earlier for we had had let it through, but it had moored and restarted after us. We cruised back past our night mooring and reached the bottom lock an hour and a half after casting off. Pearl was too charitable to mention the suggested reversing again.

The Stourbridge sixteen have a notable history and are one of the best examples of local determination and of restoration carried out by volunteers in co-operation with the British Waterways Board. In the 1962 National rally of boats the Inland Waterways Association recommended as many boats as possible to use the Stourbridge locks. It is on record that over forty boats came down the locks and that over fifty returned by that route despite a notice of danger which stated that they were unsuitable for the passage of boats. Not long after the rally finished, vandals smashed the heel post of a bottom gate, the locks were usable no more and rubbish began to be dumped in increasing quantities. We were told of someone who managed to get his boat through, working weekend after weekend and others were also successful despite the rubbish behind the gates. So, at the end of 1962, the locks were scarcely usable and apparently doomed. Nevertheless the Midlands branch of the IWA did not lose heart and further they did all in their power to keep the rubbish within bounds and to encourage boats to attempt the locks.

It was at the end of the summer of 1964 when the news was announced that British Waterways Board had invited the Staffs & Worcs Canal Society to collaborate in the restoration of the locks 'to minimum pleasurecraft standards'. Work was to begin in October and to be completed by the end of the 1965 cruising season. Actually the task was one of very hard labour and took a full three years to accomplish. A third of the gates were to be replaced by the Board assisted by the volunteers who were also to remove all the rubbish.

Right from the start, the name David Tomlinson stands out as the organiser and driving force who kept the work going and saw it to a successful conclusion. He was present every weekend, summer and winter alike even though the work took longer than was anticipated. He mobilised volunteers from the locality and from farther afield and acted as the essential link with the Board.

With their faithful narrowboat *Wallace* always at hand, the Board and the volunteers pressed on with the clearance of the canal and the towpath until on 7 May 1967, Mr John Morris, Parliamentary Secretary to the Ministry of Transport, declared it officially reopened. In his speech he commented, 'Only use will justify the money that has been spent.' This was one reason why we were so keen to use the canal ourselves.

Navigation, the magazine of the Midlands Branch of the IWA, recorded the restoration in detail and described the hard slogging work, the mud, and above all the enthusiasm and humour of the little party that worked on steadily in their spare time. Here was Stratford all over again, without the huge draglines of the army but with the Board co-operating closely. It could well be the pattern for the future, bringing back more and more miles of lost waterway into the canal system and opening up routes which we had feared were lost for ever. The time is quickly coming when we shall need all these urban and rural miles which must not be lost beyond recall.

We found the bottom lock empty and Pearl steered *Rose of Sharon* into it. The paddles moved easily and we started to ascend the flight on the lovely Whit Monday morning. At first we were still in green countryside but after a few hundred yards we passed the tall glasscone of Messrs Stuart's Crystal Works of Wordsley. This brick kiln, higher and more conical than the bottle kilns of the Potteries, is said to have been built beside the canal about 1790 and to have been in use until 1939. Such cones are few in number and this example has now been scheduled as an ancient monument. It certainly dominates the landscape for some distance around.

Stourbridge has been a great centre of glassmaking for over 350 years. We were told that when James I decreed in 1615 that all glass should be made in coal-fired furnaces in order to save the woodlands from destruction, the manufacture of glass settled in areas where coal could be dug. The medieval centre of glassmaking had been the Weald but it moved in the seventeenth century to Newcastle-on-Tyne and Stourbridge. The first glass made in these areas used the local beds of sand, but later purer sands were imported and clearer glass was produced. Stourbridge was particularly attractive to glassmakers for the very high quality fireclay in the area besides coal. The pots for melting the constituents of glass—sand, potash, saltpetre and red lead with small quantities of other minerals—need to be able to stand very high temperatures without collapsing, and pure fireclays are necessary.

W. H. B. Court, in *The Rise of the Midland Industries*, describes how Stourbridge had seventeen glass houses in 1696 and, though the number had dwindled to eleven a century later, over five hundred people were still employed in the glass trade. Stourbridge became famous and has remained so for cut crystal of very high quality, though in bygone days a great quantity of brightly coloured glass was produced for the Birmingham trade.

H. J. Haden, in his paper *Notes on the Stourbridge Glass Trade*, published a map of 1774 showing the proposed route of the Stourbridge canal. A number of glass houses occurred along the route and the preamble of the Act of 1776 mentioned the glass houses which would benefit by the navigation. Glass, like pottery, breaks easily and the eighteenth century must have produced a high casualty rate. It is little wonder that the glass manufacturers would welcome the smoother carriage provided by the gentle movement of boats.

Mr W. E. C. Stuart of the famous Stourbridge glassworks has kindly sent me a sketch of the glasscone when it was working. The cone itself was a great chimney, its height assisting in the production of a strong forced draught. Set in a hole in the centre of the floor of the cone was a furnace fed from a passage beneath. Surrounding the furnace at floor level were eight to twelve crucibles known as the pots. Round the furnace and pots was a wall with gaps through which the molten glass could be extracted for the-glass blowers. Over the furnace was a cover with flues at the sides, opening to the main cone. The men worked inside the cone around the furnace area, within reach of the pots. Mr D. Hogan of Pilkington's Glass Museum tells me that hot conditions of working were necessary to prevent the glass from cooling and hardening too quickly. Such conditions reminded me of a steelworks I once visited in Wales where I was told the steelworkers used to be able to quench their thirst with free beer. During the shift they told me 'Some drink fifteen pints of beer, and others drink a lot'!

Running out from the side of the cone was the annealing furnace with the lehr, a much cooler oven, in which the glass could be kept at an even temperature and then cooled off slowly. This is necessary, for quick cooling would set up strains within the glass itself and it would easily shatter.

W. H. B. Court, in his book mentioned above, describes the highly specialised technique of glassmaking. The glassmakers worked in teams of four, each team known as a chair. Each member had his own special duty, for the time available to the blower was limited and the process required both speed and skill. Today, the glass-blowers and their colleagues are just as highly skilled but the great heat of the furnaces can be achieved by more modern methods and the ancient glasscones are no longer required. This example still stands by the canal, a witness to generations of glassmen and the splendid products they are still able to produce.

Half-way up the locks, a young man stopped to help us. He was exceedingly interested in the canal and very knowledgeable and he told us that his interest started when he went on a week's cruise from school on the narrowboat *Ernest Thomas*. We had often met this boat on the 'Shroppie' with its parties of enthusiastic boys, always enjoying themselves. This boat, and others similarly run, must be of great benefit to young people.

We continued up the locks towards a road bridge which crosses the flight near the top and has on it a huge poster showing a two-handled mug of beer. Thus encouraged, we pressed on into the penultimate lock which has beside it the lock cottage. We moored above and accepted the lock-keeper's invitation to reach the road through his garden and there we enjoyed a pint amongst friendly people who were interested to know we were using their canal. We returned to *Rose of Sharon* for lunch and then entered the top lock.

The Stourbridge canal and its neighbour the Dudley canal are both early in the history of canal building in England. In 1766, Acts of Parliament permitted the construction of both the Trent and Mersey and the Staffs & Worcs canals and two years later a further Act enabled a canal to be built to link the Staffs & Worcs canal with the important and growing industrial districts of Wolverhampton and Birmingham. What we now call the 'Black Country' was then green and hilly with coal and iron ore where lived a great variety of craftsmen ready to make use of the industrial developments of the mid-eighteenth century. Such an area was poorly served with communications and the canal proved a great boon.

There is, however, a range of hills south of the Wolverhampton to Birmingham line and a further important and coal-mining area farther to the south. These hills, little short of 1,000ft in places and including Wrens Nest, Dudley Castle, Netherton Hill and Rowley Regis, formed an effective barrier and towns to the south needed their own outlet for trade. Thus, we find the two Acts of 1776 for a canal to Stourbridge from the Staffs & Worcs with extensions northwards into the coalfield of Pensnett Chase and a canal at Dudley to link up with these extensions, recognising these needs. It is odd at first sight to find separate Acts for what were really two sections of one navigation and Hadfield tells us in *The Canals of the West Midlands* that an attempt was made earlier to have the canal built by one company, but the single Act was opposed and failed. In fact, the two companies were very different in outlook, the Stourbridge quiet, conservative and wealthy and the Dudley lively, active, promoting Act after Act and having little money to spare for the first fifty years of its existence.

The Stourbridge canal took the line we had already travelled and forked at the top of the Stourbridge sixteen locks, the left branch into Pensnett Chase no longer used. The Act itself has a number of interesting clauses including one which prohibited coal from being extracted from beneath the canal. Mines were not to be worked nearer to the canal than 12yd and tunnels only were to be driven beneath the canal. These must not be more than 6ft high and 4ft wide and not nearer each other than 9ft. If all coal-mining in the last 200 years had been bound by similar clauses, many problems of subsidence might not have had to be faced. Further, the canal was nowhere to approach within 1½ miles of the Birmingham canal, for older companies always watched their interests jealously.

As we rose in the top lock we had a splendid stroke of luck for we were soon chatting with a man and his son who were sitting on the top beam. They opened the gate for us and, as they were taking the Whit holiday, we invited them aboard for a short cruise. As we set off, turning to the right at Brockmoor, I realised that the man, Mr John Hemming, knew the canal intimately and it turned out that he used to work boats up and down the locks. He told us that tugs would bring boats to the top of the Delph locks and that he and others would have their horses ready to take them down through both the Delph and the Stourbridge flights of locks, returning later in the day with a boat from the lower level. He knew every yard of the canal and every building on it. 'That is the forge where chains were made for Nelson's flagship, the *Victory*.' 'The anchors for the *Queen Mary* were made in that building.' The whole length of canal round Brierley Hill is full of industrial history, and here was the guide we needed to describe it to us. I wished I had a tape recorder to retain his account, for we saw so much and the journey was over so soon.

At length, we came to the bottom of the Delph locks which John Hemming referred to as 'the nine'. There are, in fact, eight locks only though the Ordnance Survey map marks 'Nine Locks Works' near the bottom. Hadfield tells us that there used to be nine but in 1858, the middle seven were replaced by six. They are, in fact, the finest flight of urban locks I have ever seen.

As we reached the bottom lock, John Hemming seized a

windlass and set the paddles and I went on to the next to make it ready. As water poured in, Pearl found *Rose of Sharon* rising like a cork. I was a little concerned that we should bump the bows or the stern and remarked that in locks strange to me, I liked to take things slowly. John Hemming replied 'I am familiar with these locks' and there was a wealth of meaning in these words. They were not just 'I know them' or 'I am used to them', there was something of a oneness between the man and the locks that made for perfect operation. I think our actual time of ascent of the eight locks was only thirty-five minutes and I spent quite a lot of time taking photographs.

The bottom lock is below a slight curve which leads on to the next six. These stand one above the next, each rising between 11 and 12ft and each with a side pond whose excess water tumbles over a series of rapids into the next. I photographed the date 1858 carved into the masonry of one lock and took views up and down the flight. The older series of locks used to lie a few yards to the south-east of the present flight and the branch to them can be seen from the road bridge above the six. The top lock lies beyond the bridge round a further slight curve.

I look forward to revisiting the 'nine' which were such a pleasure to lock through on this occasion. They are, in fact, the start of the Dudley canal whose purpose was to tap the important coal and ironstone deposits lying south of the ridge of high ground but to keep well clear of the Birmingham canal. It was also to tap some of the vitally important limestone deposits which were quarried within the great dividing ridge. In 1785, a further Act repealed a clause which kept the canals apart and permitted the building of a link between the two systems, which would pass beneath the town of Dudley in a narrow tunnel, 1¾ miles long. This was to join a short canal already in existence belonging to Viscount Dudley and Ward which had been built to carry lime from his mines. This little canal linking with the Birmingham canal was to be incorporated into the new Dudley canal with a stop lock to preserve the older canal's water.

Five years later, in 1790, yet another Act allowed the raising of more money to complete the work and a more direct line was built from the north end of the tunnel.

From the top of the Delph locks, we cruised northwards towards the great mass of Dudley hill. We passed the sealed-off end of the so-called two-lock cut, a short cut across the valley to the right which had been built on an embankment. This has long been dewatered owing to subsidence over coal mines. To our right was a great steelworks. Soon we reached the last lock which lifted us up to the Birmingham level of 453ft above sea level, a deep lock which had originally been two, replaced at the same time as the Delph locks were altered. Since Stourton, we had come up twenty-nine locks which had raised us no less than 278ft.

We were now at Blowers Green junction where the line came in from Dudley tunnel. We moored and walked up the three Parkhead locks to the small, oval tunnel entrance. The locks at that time had good top gates but were full of rubbish and needed extensive repairs. We were later to have a trip through the tunnel (see page 157) and to see the great amount of work the volunteers had carried out on the locks and their surroundings.

The tunnel itself is a remarkable structure which cuts through both coal measures and limestones. These limestones, of Silurian age, were formed about 400 million years ago in shallow, tropical seas which covered much of Britain. In warm waters, corals built reefs and an immense variety of shell life abounded. Long after the coral and shell debris had collected and been compressed to form limestones, they were folded into great arches which now form the hills of Dudley Castle and Wrens Nest. Many years before I became interested in canals, I collected fossils from these hills and studied their formation and am still working on a group of small coral-like fossils which are exceedingly common.

Lime is essential in both industry and farming, and the pure limestones of the hills have been quarried for centuries. The rocks dip steeply into the hillside and the quarries have run deeper and have become mines. Where the tunnel crosses these beds of limestone large caverns open out and shafts were driven sideways into the rock. A separate tunnel, 1,200yd long was driven under Wrens Nest to help to bring out the stone. The carrying of stone in the main tunnel caused severe hold-ups to through traffic which were only alleviated when the alternative route was built through the hill at Netherton. A few years earlier there had been talk of sealing off the south end of Dudley tunnel by a railway embankment for a line which was itself threatened with closure. The Dudley Tunnel Preservation Society was formed under the chairmanship of the present Viscount Dudley and Ward and the tunnel was saved. More recently, the society, in conjunction with Graham Palmer's *Navvies*, has tackled the Parkhead locks and the area around the tunnel entrance. Furthermore, many of the members have joined together with others and, in close co-operation with Dudley Council, have formed the Black Country Museum, to be set up at the Tipton entrance of the tunnel astride Lord Ward's branch. In the autumn of 1971, a large party of volunteers cleared out the branch canal in readiness for the museum and the tunnel's future seems bright.

We took a number of photographs and returned to *Rose of Sharon* for tea. Reluctantly, we said good-bye to John Hemming and his son who had been such excellent guides and cast off to turn east for a night mooring under Netherton Hill.

The canal, which had come up through eighteenth- and nine-teenth-century factories and which had passed by twentieth-century flats and steelworks, ran now from Blowers Green round Netherton Hill along a delightful stretch of green country-side. There was a wide towpath and a tall hedge of hawthorn in bloom and the hill, once heavily scarred with coal tipping, was now greened over. Standing proudly on its summit was Netherton Church, a brown stone building with a square tower. We moored up in peaceful surroundings in deep water against the stone kerb of the towpath.

Next morning we were up early and off again through

'Brewin's Tunnel', a deep, bridged cutting. Called after Mr Brewin, an able superintendent in the early days of the canal, it had originally been a short tunnel and had been opened out. We then cruised through the highly industrial section of Netherton and past the Bumblehole loop to the final junction before the great Netherton tunnel.

Scarcely had the canal company completed and paid for the tunnel at Dudley when they had a fifth Act in 1793 permitting the building of the Dudley No 2 line. This was to be cut on a level through exceedingly hilly country by Coombs, Halesowen, Lapal and Weoley Castle to Selly Oak where it was to link with the Worcester Birmingham canal, permitted by Act of Parliament in 1791. It was to run round the contours and cut through the main ridge of hills by the immensely long Lapal tunnel—$2\frac{1}{4}$ miles long. It was also to pass through Gosty Hill by a fairly short tunnel. Though there were to be no locks other than a stop lock at Selly Oak, it was to be a difficult engineering feat and it is hardly surprising that the company obtained a sixth Act in 1796 to allow them to raise more money.

Hadfield describes fully in *The Canals of the West Midlands* and I have mentioned more briefly in *Water Rallies*, the competition which this new line would have with the Birmingham canal and the opposition that was raised. Suffice it to say here that it was completed successfully in 1798 and that it formed an important link in the route to London via the Stratford canal and its connection with the Grand Union (then the Birmingham and Warwick canal) at Kingswood, opened in 1802. Despite the length of journey below ground, we wished that we could have chosen that way. Unfortunately the great tunnel at Lapal, sometimes spelt Lappal, subsided in 1917 and it is not now possible to go beyond Coombs. Further, the Selly Oak end has been completely filled in and nothing can be seen of it from the Worcester Birmingham canal today.

At Gosty Hill, the canal runs through a fairly high ridge in a short but impressive tunnel. It enters through a small portal but opens out inside, reappearing at the eastern end from an almost vertical hillside. It then runs through a steelworks with a large fleet of narrowboats used for storing steel billets and rods and moving them about the works. They also had tugs which were used until quite recently to pull boatloads of coal from the Cannock area. Since the extension of the canal has been abandoned beyond the A5 road at Norton Canes the coalfield has been cut off and coal can travel by water no more. As so much of the line was on one level, it seems a pity that this trade has been lost. The narrowboat fleet has all the services including repair workshops and a launching slipway. We even saw a large stack of rudders ready for fitting if required.

The canal remains navigable for a further mile beyond the steelworks and on the south side is a magnificent basin with a huge shed covering railway lines, now disused, along one side. If this is not to be used again for its original purpose of trans-shipment of goods, it could be turned into a fine marina with proper fencing and security. Throughout the canal system, there is a great need for safe, secure moorings and in the populous Midlands, more and more people are buying pleasure boats. The marina at Braunston is a good example of the proper use of a large area of water connected to the canal and surrounded by all the essential services. We could only hope that this basin at Coombs would receive similar treatment and would not be abandoned and finally filled in.

Beyond the basin the canal is full of weeds and a road bridge has been dropped to make further navigation impossible. The next few miles, now cut off, are in rural surroundings. The great Lapal tunnel is now completely abandoned with tipping near the entrance. Hadfield tells us that a clever scheme was in force from 1841 to 1914 to help leggers to pass the boats through. This originally took up to four hours but a pump was installed to cause a slight current to flow through in one or other direction depending on the way the boats were to be legged. Nevertheless, the 2¼ miles of darkness must have seemed endless to the leggers even with a slight flow to help them.

Our cruise from the junction with the No 2 line took us towards Netherton tunnel past an old pump-house which had once lifted water from the coal mines. Standing starkly on the hillside with its tall brick chimney, it was one of the 'fire-engines' we read of in the canal Acts. These pumps, normally worked by beam engines, made it possible for coal to be followed to much greater depths. The water, sucked out by the pumps, was a valuable addition to the canals which needed replenishment as the boats worked through the locks.

Netherton tunnel is quite unlike any of the other long tunnels in the area. It was the last to be built and was completed in 1858, the same date as the rebuilding of the Delph locks. It was the widest canal tunnel at 27ft and has towpaths running along both sides. Originally lighted by gas and later by electricity, it was now in darkness and we needed our headlight. Before we entered, we could see the other end clearly and we ran through easily and without trouble. There are a number of ventilating shafts most of which drip like taps and I needed my oilskins and sou'wester. I glanced up two of them and saw points of light 400ft above me and, though we were on the 453ft level, we felt as though we were deep in the ground.

Beyond the tunnel a short, straight length leads to the Birmingham main line. We passed beneath an aqueduct carrying the old main line on the Woverhampton 473ft level and turned right to make for Birmingham.

Throughout the rest of the journey to Worcester Bar, we were on Telford's straightened section, a wide, deep canal, with towpaths on both sides, often lying in deep cuttings. We passed beneath the slender, cast-iron Galton bridge and later the short tunnel which carries Broad Street, with its shops and little church, over the canal.

We had travelled the 'Black Country' from side to side. Our chief impressions were of the wonderful canal structures, the great tunnels, the well-designed locks, the cast-iron bridges and many other eighteenth- and nineteenth-century monuments. We were delighted with the little green corners of countryside which often lie so close to the mines, the foundries and the forges. And we met and talked to a lot of warm-hearted and interesting people. It is a miracle that some of these waterways are still open and navigable. Now, with the co-operation of the Board, Local Authorities and a host of volunteers, they look forward to a bright future.

Page 51   (*above*) Snareston tunnel on the Ashby Canal; (*below*) Coventry basin
from the air during a boat rally

Page 52   Erewash Canal: (*above*) Trent lock, above the junction with the Trent; (*below*) Sandiacre showing the lace mill with its distinctive chimney

CHAPTER THREE

THE ASHBY CANAL

WE had passed the entrance of the Ashby canal many times when cruising along the Coventry en route for the Thames. On each occasion we determined to make time to explore this peaceful little waterway to its limit. Our opportunity came when we had been down the Grand Union to Stoke Bruerne and were on our way back home. After all the heavy locking, for we had come down through Warwick, we welcomed a level canal without a single lock in the whole of its 22 miles. Originally it had extended a further 8 miles to Moira and had still been lockfree.

We turned in through Marston junction between Bedworth and Nuneaton, past a faded blue notice which read 'Ashby Canal Only'. The notice was clearly meant to warn people that the canal did not link with any other and that they would have to return the same way, but it hardly seemed to invite exploration of this lovely little navigation. Through the red brick bridge is an old stop lock with ancient wooden gates now left permanently open for the level of the two canals is exactly the same. Beyond this, the channel turned out to be surprisingly deep and clear as it was then kept open by the regular passage of coal boats.

The Ashby canal owed its birth to coal as it was built to carry coal from the Leicestershire mines around Measham and Moira. It also owes the loss of the top 8 miles to coal for the mining in these areas has led to steady and persistent subsidence

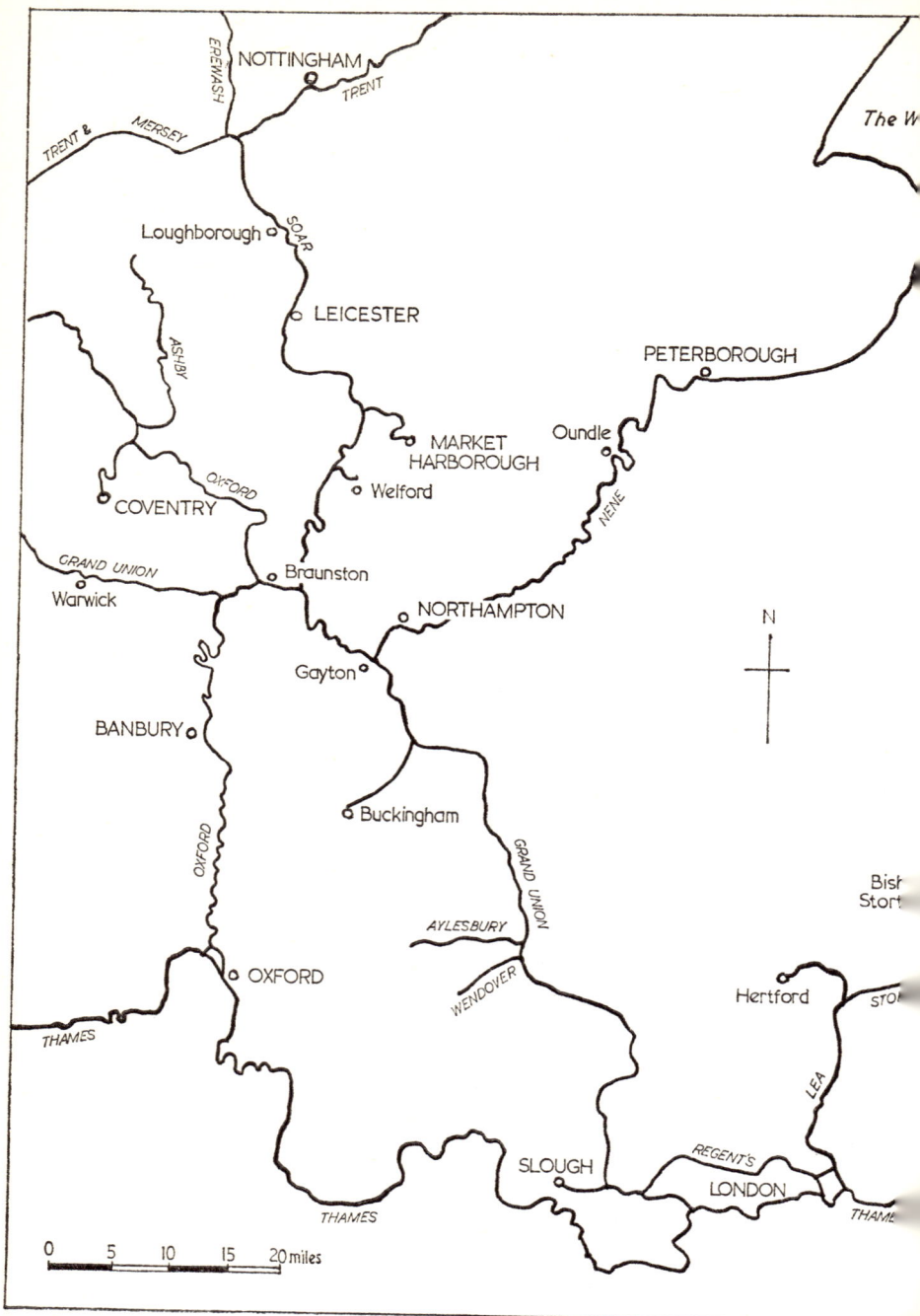

Midland and southern canals and navigable rivers

which has caused the abandonment, almost mile by mile of the top 8 miles. Thus the canal, which had started as an outlet for the coal mines, now stops at the edge of the coalfield and the coal must be brought to its banks by road. For the pleasure boatman, however, the 22 miles are in rural and not mining country, well beyond all possible subsidence.

Looking at the canal map of England, one remarkable unfinished line seems to stand out. A short distance north of the original terminus at Moira, the ground rises and then falls away to the Trent valley at Burton. Furthermore, the river Dove enters the Trent from the north-west and 10 miles up the Dove valley, at Uttoxeter, a branch of the Caldon canal used to terminate. Had a Bill of 1796 become an Act of Parliament, the top pound of the Caldon canal might have been linked with the Macclesfield. Such a line would have produced a far more direct route from Manchester to London than the earlier Bridgewater, Trent and Mersey and Coventry canals route. Needless to say, such links had their supporters at different times and Hadfield in *The Canals of the East Midlands* has described the various schemes, together with the histories of the canals that were constructed.

The Ashby canal was first mooted as early as 1781 as a link between Burton-on-Trent and the Coventry canal, a through route which would transect the little coalfield and tap the limestones of Breedon and Ticknall. This met with opposition from the Coventry canal company as it would have competed with their own line to the Trent and Mersey at Fradley, though that had not then been completed. An alternative plan to link the coalfield of Ashby Woulds to the Coventry received the latter company's blessing. Nothing was actually done and the schemes were revived on several occasions during the next twelve years. Finally, after considerable difficulties, an Act was passed permitting 50 miles of canal, 30 on the level from Ashby Woulds to Marston, a mile east of Griff on the Coventry canal and the remainder forming a series of branches from the various limestone, ironstone and coal-mining areas. These branches would have crossed the main watershed to the north and could easily have been linked with the Trent at a later date.

The level section was opened in 1804 but the cost was much greater than had been estimated and it was decided not to build the branches but to lay out tramways instead and this was accomplished successfully. Unfortunately the little coalfield was not as productive as had been hoped and it was many years before the shareholders received any dividend. At length more pits around Moira were dug and were found to produce a very good quality of coal which sold well in London and the income from tolls began to rise.

In 1796 the idea of the Commercial Canal was mooted. This was to have been a wide canal from the Chester canal at Nantwich, the Dove Valley to Burton-on-Trent, thence to the Ashby canal, the Coventry and Oxford to the wide Grand Union and London. The scheme was defeated the following year but once more it would have made the Ashby a through route. The idea of the through route was still alive even in 1840 with a plan for a Burton and Moira canal, but by then the railway age had started and the opportunity was lost.

In this century subsidence in the Moira area was becoming serious and the top $2\frac{1}{2}$ miles was abandoned in 1944. Nearly five more miles went in 1957 and Ilott Wharf, Measham, was cut off and dewatered in 1966. Thus the little coalfield swallowed up its canal links one by one and the waterway now stops short of the coal measures.

The Act of Parliament of 1794 contains most of the usual clauses, many of which are of rather special interest. A lengthy section is devoted to a spring which supplied a reservoir in Gopsall Park from which Penn Assheton Curzon of Gopsall House (not Gopsall Hall) drew a quantity of water. Curzon had only agreed to the canal being built on the understanding that this should not be adulterated or diminished. If the spring was affected within four years of the canal being built the company was liable to pay £50,000 compensation! Fortunately they were never required to do so.

The company was permitted to build boat rollers or inclined planes instead of locks in the hilly northern section. The coal mines near the canal were to pump water into it with their 'fire-engines', mostly beam engines operating pumps, and none was to mine coal beneath the canal without due notice being given to the company. If only they had been forced to maintain a pillar of coal beneath the canal or to follow the rules of the Stourbridge and Dudley canals which permitted no coal to be mined beneath them and small connecting tunnels only to be built, the subsidence would not have occurred and we should still have had the full level 30 miles today.

There is an interesting clause on the preserving of the land on the canal side. Any clay, gravel, sand, rubbish or other materials (not soil) not used in the construction of embankments was to be spread evenly over hollows, but first, 9in of soil was to be lifted from the ground, kept quite separate, and replaced over the new ground surface. We know this is done by large earth-moving machinery today and it is interesting to see the importance attached to it in 1794 when 'navvies' dug it all out with spades.

The canal was to have a stone or post clearly marked every quarter of a mile so that all would know where they were. It was realised that much of the carriage would be in one direction only, from the mines and quarries southwards, and encouragement was given to carriers by permitting toll-free return journeys for empty boats.

The usual paragraph on identification is very specific—the owner to paint his name, abode and boat number on either the head or the stern of the boat in white letters at least 6in high on a black ground above the level to which a laden boat might sink. Landowners might use pleasure boats without charge but could only use locks when excess water was flowing over the weirs. As the only section built was level throughout land-owners had full free use of the whole navigation providing their pleasure boats carried no goods or persons for hire. The land-owners also had the fishing rights.

It is interesting to note in clause 113 the varieties of goods the boats might be expected to carry. These are listed as 'ironstone, coals, lime, goods, merchandise, commodities, matters or things whatsoever'!

The last clauses are ones of protection to the earlier built Coventry canal and the permitted, but not yet completed, parallel Leicester canal. Each of these companies was empowered to erect a toll house and a bar. The one was to be on the south end to collect tolls for boats passing out on the Coventry canal (the toll from Marston Bridge eastwards to be reckoned as though the junction had been at Griff!) and the other at the northern end to collect compensation tolls for the carriage of coals from the Swannington-Coleorton district which might otherwise have found their way on to the Leicester canal. It was no wonder, with such restrictions and the fact that the canal was never extended northwards to the Trent, that many years were to elapse before the company could pay a dividend.

As we cruised up the little Ashby canal we regretted the loss of the top 8 miles and wished that we could have gone to its original summit and continued on northwards to the Trent.

The waterway winds around the contours of the hillside and apart from a narrow rock cutting on the edge of Nuneaton, a mile from the junction, it is at field level for most of its course. What struck us most was the wealth of flowers which included the small water lily, king cups and huge clumps of the golden flag iris on the edge of the towpath. We cruised under the busy A5 Watling Street and remembered the same road both at Gailey lock on the Staffs & Worcs canal and running beneath the aqueduct on the 'Shroppie' near Brewood.

The only arm left is at Hinckley where a 300yd cut runs eastwards to the old town wharf which now provides moorings for pleasure cruisers. We were short of supplies but decided not to stop at Hinckley but to push on northwards. From here onwards we were in very remote countryside and were soon looking for a good mooring for the night. Though the channel was deep, the sides had been very shallow most of the way and we had some difficulty finding a place near enough to the towpath to land easily and to be out of the way of any narrowboatmen moving late or early. We saw a few 'temporary moorings' marked near the village of Stoke Golding but they were occupied by seven small cruisers which looked very permanent. We finally chose the towpath near to Dadlington and walked up into the pleasant country village.

Next morning we were up early and my log notes 'very pleasant cruising through fields and parkland'. Our 1in map told us that we were passing through Bosworth Field, the scene of one of the most decisive battles in history. Around us were probably the descendants of the thorn bush from which Richard III's crown was plucked to be presented to Henry Tudor. Very few battles in history can really be said to mark the end of a chapter and the beginning of another. Hastings and Waterloo are both in this category and so is Bosworth Field. Here, we are told, Richard III the last of the medieval kings, fought fiercely and bravely to be beaten and killed by the soldiers of Henry Tudor marching under the Red Dragon of Cadwallader. To this day we are not sure whether Richard was a demon and murderer from whose grip the country was saved by the romantic Welshman, or whether Richard, a good and able ruler, was tricked by his enemies who were led by a usurper with a trumped-up claim to the throne. We do not even know if Richard was really a hunchback at all. As we cruised through those peaceful fields with early wild roses in bloom and the thorn still carrying May blossom we learnt no more of the protagonists and 1485 seemed a very long time ago.

A few of the bridges only bear numbers but bridge 40 was one which we particularly noticed. It was on a sharp left-hand bend and we were half-way through when we first saw the bows of a narrowboat loaded with coal. Thinking to give him the deeper outside curve, we cut in on the wrong side and ran firmly aground. It took a certain amount of poling after he had passed to get us free.

Market Bosworth wharf is the only water-point on the canal.

The wharf itself is formed by the sinking of two iron narrow-boats, now filled with soil, which make a good mooring. The map showed the village to be some distance away so we cruised on through rather more weed than before, though nowhere bad enough to hold us up.

Another hour's run brought us to the little village of Shacker-stone. Having crossed the river Sence and passed one of the two bridges, we cruised past a small field with what looked to be an earthwork with a tree on top. Beyond the second bridge we reached an excellent mooring opposite a wide winding hole and from there we were able to walk back into the village. As we passed the church the clock stood at midday.

Wanting to collect stores, we inquired for the village shop and were directed to the post office but were warned that 'they shut for lunch'. The office itself proved to be a delightful cottage, but it had already closed at midday. We retired to the inn, a delight-ful free house which had not long been in the hands of its present owners who told us that they had always wanted to run a country pub. After lunch on the boat we pushed on northwards.

For about a mile the canal skirts the woodlands of Gopsall Park and this had a special interest for us. Pearl's maiden name is Jennings and there was supposed to be a remote family con-nection with Charles Jennens who, two centuries ago, enter-tained Handel at Gopsall Hall where he is said to have composed 'Messiah' in the incredibly short space of three weeks. We had been told in Shackerstone that the Hall was occupied by the military during the war and that it had now been demolished. The park was obviously well cared for and the woodlands were very beautiful. Even today it would be a wonderful retreat for anyone who wanted complete quiet to compose a great work and 200 years ago it must have suited Handel's purpose admir-ably for even the slow canal traffic had not yet begun. I told Pearl that I approved of her ancestral home!

Beyond Gopsall Park we came to the little Gopsall wharf from which coal was loaded into narrowboats. From 1965, for a short time, the boats had been able to load at Ilott wharf near

Measham where electric equipment was available and there was space to moor and turn. This had been lost in the last short closure though the board had made a winding hole at the extreme northern end of the navigation section. Rather than lose the carriage of coal the Ashby Canal Preservation Society had prepared Gopsall wharf and its approaches and had made them available for loading. Lorries then carried coal to the wharf to be shovelled on to a small belt loader which tipped it into the boats which had had to journey to the extremity of the canal to turn round.

Beyond the wharf we reached Snarestone and moored for shopping. Once more we inquired for the shop and were told that it was the post office. This had closed at one and again we were unlucky. However, it was a good excuse to visit Measham only 2 miles to the north which we could have reached by boat a few years previously. My sister had joined us for this part of the trip and she inquired for and caught a bus. I missed it by seconds and set off to walk. We arrived almost simultaneously for the bus had taken her on a pleasant tour of the countryside and neighbouring villages and had covered at least twice my distance. As I entered Measham, I passed by a fair which had been set up on a field adjoining the now filled-in wharf.

We did our shopping and then looked to see if we could find any pieces of that much sought after boatman's Measham pottery. This deep brown earthenware with flowery patterns in slip is famous throughout the country. Many of the examples have short sentences which may include the owner's name and possibly the date and the most prized pieces are huge teapots, occasionally with two spouts, crowned with a lid resembling a small teapot. We were unlucky for there was nothing to be found and it surprised me that a small mining town far from the more famous pottery areas should have given its name to pieces of such distinction. I determined to find out more about the history of this trade and this proved more difficult than I had expected. I first went to the pottery reference books in the library but found no mention whatever of Measham or pottery made in that area. As a museum director, I am fortunate in having contacts with specialists in most subjects and I phoned up my colleague Arnold Mountford, director of the Stoke-on-Trent museum whose pottery collections are world famous. He surprised me by telling me that very little is written on this distinctive ware and that his own notes were few and sketchy. He had recently sent them to Mrs Griselda Lewis who was kind enough to send me the scanty information that she had then collected. She told me that marked pieces were known but that she had not yet seen one and that several pieces had dates in their inscriptions. The earliest she had seen was 1881 and the latest a mere twenty years later. Unfortunately most of the evidence was hearsay—certain pieces were said to have been made by such a firm in such a place.

They would definitely appear to have been made at Church Gresley and Woodville, both suburbs of Swadlingcote, the neighbouring town to Measham. Rolt mentions in *Narrow Boat* that an old lady on the Moira cut sold these pots to the boat people but had stopped doing so about 1910. Moira itself is within a couple of miles of Church Gresley and is the nearest place on the Ashby canal to these potteries. Mountford has evidence that this ware was also made at Newcastle-under-Lyme but has not yet traced the pottery, and Burton-on-Trent is another suggested place of manufacture. All these towns are on or close to canals.

No doubt marked pieces and documentary evidence for these potteries will turn up for such pieces were still being made within living memory. Meanwhile the mystery surrounding these remarkable pots makes them the more intriguing.

We rejoined *Rose of Sharon* moored by the little Snareston tunnel. Snarestone is remarkable for a canal village—it is nearly impossible to see the canal. Though the tunnel is only 250yd long, it passes beneath the village and keeps the little waterway completely out of sight. Even the track from the towpath runs through a Waterways cottage garden and on to the road by a side gate. We nearly missed it on our return. About a week previously, we had heard that my brother was going to be in the area and that he had chosen Snarestone from the map as a suitable place to look for us. Luckily we had arranged a different rendezvous or he would never have found us.

Back on board, we cruised through the tunnel, amazingly crooked for one so short, and carried on northwards to the present terminus beyond two small accommodation bridges. The canal was stanked off and the winding hole cut in the east bank. The bed beyond was heavily overgrown but still in good condition and some distance ahead was the chimney of a pumping house. Some years ago, Robert Aickman pointed to the Ashby canal as an example of how abandonment of one length is often a prelude to the abandonment of further sections. We must see that the loss of Ilott wharf, the third section to be dewatered, is not followed by any further encroachment on this lovely little canal.

We turned reluctantly and cruised back through the tunnel, mooring for the night at Shackerstone opposite the winding hole, and walked once more into the village. This area of Leicestershire is one of strange delightful names which sound as remote as the villages themselves: Shackerstone and Snarestone, Sheepy Parva and Barton-in-the-Beans, Ashby-de-la-Zouch and Odstone, and many others as unusual. Many bear evidence of the Scandinavian invasions by Norsemen, Danes and Frisians. Ekwall tells us in his *Oxford Dictionary of English Place Names*, that Odstone is Odd's Tun, Odd being a Norse name and his tun was his enclosed homestead or village. Bilstone was Bild's Tun, Bild being a Danish personal name, and Freizeland, another village name on the map, is land occupied by the Frisians.

Several of the names bear evidence of the type of land that existed in the Middle Ages. Barton was a piece of fertile farmland, possibly a monastic grange, and beans, like peas, have always been an important article of food. Appleby Magna was a place of apple orchards and Sheepy Parva a small island where sheep could graze. The island was not in a lake or bog but in an extensive heath as two villages—Normanby-le-Heath and Heather—indicate. Measham was cultivated ground on an un-drained moss. Ashby-de-la-Zouch is simpler than it sounds, for Ashby was a clearing in the ash groves belonging to the Zouch family. We know that England was heavily forested in the Middle Ages and that primeval forests were mostly of ash trees. Oaks were also common, the acorn being an important pig food and this is recorded in Market Bosworth, the market of the enclosure for boars.

These names give a very clear picture of Leicestershire in by-gone days, an area heavily wooded with heaths on the drier ground and mosses in the low-lying hollows. It was cut by a Roman road as is shown by the names Stretton-en-le-Field and Stretton Bridge west of Measham and was occupied and culti-vated by Vikings and Norse settlers. It was later fought over by Richard III and Henry VII and then left very much to its peace-ful rural industries and farming. The canal came at the end of the eighteenth century and the little coalfield developed on the northern part of the area. Today it is still predominantly rural countryside.

There are a few names I have not mentioned—Shackerstone, Snarestone and Swepstone, near Measham. Once more Ekwall suggests the origins. Shackerstone was the robbers' village and perhaps, had we known this, we should not have slept so peace-fully at its little mooring within sight of the medieval earthwork! Many centuries must have elapsed since the robbers last returned to their village from the ash forests where they may have roamed like Robin Hood and the old Roman street where they may have lain in wait for travellers. Snarestone might have filled us with even greater apprehension for Ekwall tells us that the old names were Snarkston and Snargston and that the first syllable was a personal name. Was this where the Bellman, the Beaver and their many shipmates came to hunt Lewis Carroll's Snark? Was their epic voyage on a canal after all and were they amongst the earliest canal travellers? Their map was a 'perfect and absolute blank' so we may never know! Should we have come upon this creature in the tunnel and have found it to be a Boojam we might have 'softly and silently melted away'!

Swepstone was Sweppi's tun and we wondered if this gentleman was the first to make a certain famous mineral water. If we had had time we should have sampled it and should have expected to be served with glass bottles sealed with marbles. Alas! these three names lead us only into the realms of imagination and fantasy.

Next day we cruised back to Marston junction and left the Ashby canal. It was a brilliantly fine Sunday morning and we were wakened early by the fishermen. Our trip south was peaceful and uneventful though we had difficulty in finding places to moor close to the bank for our lunch and tea. We passed a pair of empty boats going up for their loads of coal and both we and the fishermen were grateful to them for keeping the channel clear by their regular passage.

We were cruising in mid-June which is a relatively quiet time of year but we were surprised to see so few pleasure boats. We had not met one on the previous two days and now all we saw were occasional small boats which had moved but a mile or two from their moorings and only met one pleasure cruiser from a different canal. In a way we were grateful for it must soon attract numerous boats to its long lock-free pound; nevertheless there is plenty of room for all on its quiet waters.

We finally moored for the night on the Coventry canal beyond Bedworth and were settling down after supper when Lindy fell overboard. Furthermore, despite our calls, she swam across to the other side and stood miserably in about 6in of water. We had to cast off and pole the bows across and I rolled up my trousers and went overboard to collect her.

So ended our pleasant Ashby trip, we vowed the first of many. On the morrow we were to visit Coventry and that proved to be a very different story.

CHAPTER FOUR

THE COVENTRY ARM

MOST canal pleasure boatmen know the Coventry canal. Many have used it as the through route from north to south, the least hilly route with the smallest number of locks. Planned to form part of one of the arms of Brindley's grand cross, it was to be the northern part of the link between the east to west Trent & Mersey canal and the river Thames. I noted something of the history of the navigation in *Water Highways* and Hadfield goes into this more fully in *The Canals of East Midlands*. Briefly, it was permitted by Act of Parliament in 1768, two years only after the Trent & Mersey and the Staffs & Worcs canals had received their Acts. Its engineer was at first Brindley and in three years it had been completed on the level to Atherstone but had used up all the authorised capital. It was not until 1790 that the Coventry canal was completed to Fradley and it would never have reached this far had not the Trent & Mersey company constructed the $5\frac{1}{2}$ miles from Fradley southwards and the Birmingham company the next $5\frac{1}{2}$ miles on to Fazeley.

In consequence the journey south from Fradley is on a canal built by three companies and the differences in bridge construction and other works is very marked. The Trent & Mersey company sold their length to the Coventry canal but the Birmingham company retained theirs throughout.

Most people cruising south will have turned left at Hawkesbury Junction, swinging completely round through two right angles, into the stop lock of the Oxford canal. In recent years comparatively few pleasure cruisers have passed the junction and continued straight on to Coventry. We decided to make the trip. We stayed for the night within sight of the Navigation Inn at Bedworth, a delightful place with a pleasant garden and canal-side moorings. Next morning we set off again southwards.

At the end of the long cutting we passed the little arm to the Bedworth coal mines and continued up to Hawkesbury Junction. Moored there we saw first the narrowboat *Pearl Hyde*, owned by a trust and used for taking people trips along the canal. This was called after a wonderful lady who was Lord Mayor of Coventry in 1957 and who did all in her power to improve the Coventry canal. She received a rally of boats in this city and did much to combat the views of many of her colleagues who wished to close and fill in the city's length. We took this to be a good omen. There was a group of commercial boats moored at the junction and just beyond it we saw *Friendship* and had a word with the Skinners. Joe Skinner was a 'Number One', a narrowboatman who owned and traded with his own boat and is one of the last survivors of these independent people. He and his wife have now retired and they live on their butty, but almost always appear at the national rallies of boats which are held yearly in different centres.

Our first 2 miles along the Coventry arm were uneventful. The water was deep, the weeds few and the rubbish negligible. We passed a works making concrete slabs and then ran by the gasworks. Both had fine wharfs but neither had been used for years. It seemed odd to me that a gasworks using coal and situated on a canal side a few miles from a flourishing coalfield, also on the canal side, should no longer receive its coal by water. No doubt there are plenty of good reasons for this but all could be answered if there was a will to do so. Perhaps the coal reaches the gasworks by rail. Perhaps it goes the easiest way, on large lorries which make driving in this country so unpleasant particularly when the roads are wet.

We rounded a corner and the excellent state of the first 2

miles was explained—British Waterways workmen were dredging. Their dredger and boats were in the middle of the channel
and we pulled in to give them time to make room for us to go
by. 'You'll have a job to get much further,' said one. 'What's
your draught?' said another. 'You should get through,' said a
third. We meant to.

For the next ½ mile the side of the canal had a heavy growth
of weed but the channel was deep. The water was crystal clear
and Pearl, at the bows, watched out for hazards. We approached
a railway bridge where the line appeared to be abandoned and
there, in the centre of the channel, stood a most menacing obstruction. Four narrow, angled lengths of iron stuck several feet
out of the water and there were obviously other parts of the
frame submerged. We pulled in to the towpath to take stock.
There was a solitary fisherman close by who warned us to turn
back—'Those things are iron,' he said. 'Our boat is steel,' we
replied.

First we tried to pull *Rose of Sharon* gently through by the
towpath. Our efforts were foiled by a sunken spring mattress and
a mowing machine. Then we considered doing half an hour's
work removing the obstructions but decided to leave them for
the dredger. So we reversed and then tried to nose our way past
the bars on the side away from the towpath. Fortunately the
canal was only very slightly restricted in width by the bridge,
so we went into neutral and poled slowly forward. There was a
certain amount of scraping and grating but we slid through. I
would never have decided to do this in the first *Rose of Sharon*
which was built of plywood.

We turned a corner into a long straight stretch in a slight
cutting. There was a housing estate on one side and a railway
line on the other though both were hidden for much of the way
by May trees in bloom. It could have been a very beautiful little
length if it had been cared for and the towpath could have been
a quiet peaceful walk despite the proximity to the town centre.
Actually it had an air of hopelessness and dereliction for the
reeds lined the banks and restricted the channel to a few feet in

width and rubbish was dumped all along on both sides. British Waterways workmen would no doubt improve the channel when the dredger reached the area, but it is the townspeople themselves who must resolve to save the little canal. With a lead from the city, or from a lively civic society, the canal at this point could be a joy to the people who live within reach of it instead of an eyesore. We chugged slowly along, thinking what might be.

At the end of the cutting we really reached the town. First we came to a small, well-built basin on the left with a little bridge over its entrance but instead of boats, it had a couple of abandoned cars in it. Many people have a curious idea that once something is dropped into water it vanishes for ever and that canals are therefore the ideal places for dumping rubbish which, if left undisturbed, soon attracts more. Thus an unused and uncared for canal can soon become unnavigable.

The canal then turned westwards passing through a heavily built-up area. Pearl called out to me to steer away from the towpath to avoid a large submerged armchair and we carried on. We passed Courtaulds and an important chemicals factory and then the East Midlands Electricity Board. We had noticed with amusement that the Gas Board premises we had passed earlier was West Midlands! I suppose that Coventry is just about middle Midlands. All these places *might* use the canal but all had turned their backs upon it and much of this must have been due to the uncertainty of its future and the poor condition of many stretches.

We passed the famous Coventry Climax works which had a whole row of pipes down the wall pouring water into the canal. No doubt this is water they have already extracted for cooling and other purposes. It seemed to be quite clean and the presence of fishermen on the bank showed that the fish were unaffected. One of the pipes pointed outwards instead of downwards and showered its water well into the middle of the channel so we shut our windows before reaching it.

We recalled a Coventry canal Act of 1819 with a clause that

E

gave us cause for amusement. It noted that the sides and bottom
of the canal had been damaged and persons on the towpath
inconvenienced by the washing of sheep on the towpath side, by
leading stallions and driving cattle thereon and by 'persons
bathing in and exposing themselves on the bank of the said
canal'! These were considered offences punishable by a fine
not exceeding forty shillings, or failing payment of the fine, up
to a month in gaol with hard labour—a heavy punishment for
a quick dip on a hot day. However, the canal within the city,
despite the Coventry Climax shower, did not tempt us to take a
chance!

By now the canal was wide and the channel deep. It was well
constructed and in excellent navigable order. Furthermore we
realised our nearness to the city centre for we had glimpses of
the cathedral between the houses. At this point the canal is
raised slightly above ground level to the south and east.

We turned a last corner and ran past a patch of ground which
was being cleared and redeveloped and then passed through a
tiny brick bridge into the basin itself. We nosed past an old
canal building partly occupied by a joiner's shop and partly by
the Coventry Canal Society and drew up to the wharf which was
a splendid mooring. Raised slightly above the adjacent ground,
it enabled us to look straight across to the cathedral and the new
city a few hundred yards away. The British Waterways repre-
sentative allowed us to use his own cottage gate to reach the
town and assured us that *Rose of Sharon* would lie in a safe
mooring. He had a large Alsatian dog whose presence would be
a powerful deterrent to mischief-makers.

It was lunchtime and we decided to stay aboard. Almost immediately there was a knock on the cabin top and we were
welcomed by a member of the Coventry Canal Society who had
a narrowboat moored close by. Apparently strangers were relatively rare and we were cordially received. It did not strike us
at the time that our visit was just 200 years after the passing of
the Coventry Canal Act. After lunch we set out to explore
Coventry.

We had known the city before the war and also knew something of the devastation caused by the bombing but had not passed through it since. We had seen pictures of the cathedral and were thrilled with the whole conception, finding that it exceeded expectations. We were not very impressed with the treatment of the old cathedral ruins which had an exhibition of modern art inside and various booths selling catalogues and picture postcards. This seemed too commercial. It was the city itself, however, which really surprised us with its excellent shopping centre, its fine buildings, well-laid-out streets and superb international swimming pool. It reminded us of Cologne which has likewise risen so nobly from the ashes though, of course, there was no river Rhine. I do not suggest that the little Coventry canal could ever take the place of a great river but it could still be a great asset to the city if properly opened. Coventry is within reach of most of the inland boats in the country and it would undoubtedly make a goal for any cruise if the last few miles were clear and attractive. Visitors would come in thousands if the moorings were developed and I felt that the city was losing an important opportunity. I felt so strongly about it that I wrote to the lord mayor and had a most courteous reply. I learnt in the autumn that our friends, Mr and Mrs Buckley who lived on *Valkyrie*, a sister ship to *Rose of Sharon*, that they had also cruised into Coventry that spring. They, too, had had their difficulties and had been disappointed with the state of the canal. They told us that they had felt as strongly as we had and they had written to the town clerk. Perhaps before these words are in print Coventry will have replaced itself firmly on the inland waterways map and will be a popular centre for boats.

It was a very hot day so after visiting the cathedral I went to the baths and enjoyed a refreshing swim. Reluctantly, at half-past three, we returned to *Rose of Sharon* and cast off.

Our return journey was not without its problems though we now knew where to expect the hazards. The Coventry Climax squirting pipe was regulated by someone in the factory who saw us in time and realised we were not a mirage! When we reached the long, straight, narrow cutting we rolled up on to one of the largest oil drums I have ever seen. Pearl saw it but we were well and truly on top before we could do anything. *Rose of Sharon* is flat-bottomed so she rolled back off it again having flattened it slightly and there was just room to pole past and continue. We nosed through the obstruction beneath the railway bridge, this time gently under power and were soon back on the dredged part cruising along merrily.

About a mile before Hawkesbury is Longford bridge where the original junction with the Oxford canal was made, the two navigations lying side by side for a mile. This meant that coal for the south had two more miles to travel and, as tolls were based on mileage, both companies benefitted. It was eighteen years before the two companies could agree to a more logical junction at Hawkesbury though this still included the sharp horseshoe bend.

At Hawkesbury, Joe Skinner asked us if we had had a good trip and we said we had managed it successfully. He commented, 'It's all right if you take it gently.' At half-past six we were moored again at the Navigation at Bedworth and later cruised on into the quiet countryside by the Griff Arm. We shall return to Coventry by water as long as the canal is open. We should not have dared to complete the cruise in anything but a very strong boat, but perhaps next time we shall find the navigation clear, open and landscaped and that the city will have a welcome for boats.

We were told later in the season that a group of volunteers had been working to improve the length and keep the channel clear. Perhaps this is the start of a great movement which will cleanse the whole canal in the city and make it an attractive route to the fine city moorings.

CHAPTER FIVE

THE EREWASH CANAL

TRENT lock and the entrance to the Erewash lie on the north side of the river Trent at a multiple junction. Travelling eastwards down river from Sawley, the Cranfleet cut, the route to Nottingham, Newark and the sea, lies ahead to the left. To the right of this is a short wide stretch of river with a large impressive weir. Farther to the right the route to Loughborough, Leicester, and even London via the river Soar flows in from the south. On the left, however, is the well-proportioned stone-built Trent lock with moorings and steps below and 'The Fisherman's Rest' on the right-hand side.

The Erewash canal had attracted us for several years, partly because of the splendid efforts of the Erewash Canal Preservation and Development Association and the local branch of the IWA to save it from abandonment and partly for its historical interest. Classed as a 'Remainder Waterway' except for its first mile, it runs through D. H. Lawrence's coal-mining country and now leads nowhere. In bygone days it formed part of an important series of canals serving the many industries on the borders of Derbyshire and Notts. At its northern end it was joined by the Cromford canal which ran past Arkwright's famous mill and the Butterley ironworks. At the Langley Mill junction with the Cromford canal it was possible to turn south-east along the Nottingham canal into that city. Half-way along its length was the short Nutbrook canal, recently the subject of a book by

Peter Stevenson, and a little farther south the Derby canal ran westwards through that town and on to join the Trent & Mersey at Swarkestone. Now all are abandoned except the Erewash.

The canal itself had paid one of the highest dividends of any navigation and was of great commercial success. This was because it was built at an early date before the rise in costs and wages during the Napoleonic wars and was therefore relatively cheap to build. Its Act of Parliament of 1777 permitted the raising of £15,400 and a further £7,700 if necessary, and the actual cost was only about £21,000, or less than £2,000 a mile. It was open to Ilkeston in April 1779 and throughout its whole length to Langley Mill only a few months later. Even so, it was not a great success at first until more mines were dug, but was paying a 20 per cent dividend in 1787 and 30 per cent in 1794.

In the second half of the nineteenth century the canal and its neighbours lost a great deal of trade to the railways, though Hadfield tells us that it was still declaring a dividend of 11 per cent in 1870. In 1931 the Grand Junction (now Grand Union) bought the Erewash as well as the Leicester and Loughborough canals but commercial traffic came to an end in 1952. In 1962 the Langley Mill to Ilkeston length was legally abandoned, though it is still open for cruising.

We had a couple of days in hand on our way via Leicester to the Northampton boat rally and this gave us the opportunity we needed. Furthermore we knew that all boats which cruised to the summit at Langley Mill would receive a certificate to confirm the trip, so we turned into Trent lock and moored above it. It was Thursday at lunchtime—the right moment to inspect 'The Fisherman's Rest'. In this attractive pub there is a painting of Trent lock in the early days of this century and it is interesting to note the difference from the present day. Of the buildings shown on both sides of the lock, most of those on the west were destroyed by bombing during the war. In place of the concrete steps below the lock were grassy banks.

Of the canal itself, the first thing we noted was the covering of duckweed over much of the surface. *Rose of Sharon*'s air-cooled engine took no notice of it and we set off northwards. We first passed groups of houseboats and then cruised by a marina in a large basin. After a mile, we reached Long Eaton with very spectacular lace mills, large red-brick buildings whose tall chimney-tops were ornamented. Beyond the town centre, long rows of small houses backed on to the canal. Each had a long garden and almost without exception the gardens were well kept and pretty, and made a special feature of the canal bank. White water lilies grew in the shallow water and flowers were in profusion. We had never seen a canal-side treated so successfully by town-dwellers.

Lacemaking in Long Eaton has been an extremely important though relatively recent industry. The excellent pamphlet, *A History of Long Eaton*, compiled by J. E. Heath and his group of WEA members and published by the Long Eaton Council, describes how it came to the town. The hosiery manufacturers of Leicester, Nottingham and Derby turned to the production of machine-made lace when suitable machines were invented and developed in the early nineteenth century. The industry was mainly centred on Nottingham for the first half of that century. In the 1840s a number of small lacemaking factories were built in Long Eaton mostly in yards some distance from the canal. In the next thirty years the population grew from 859 to 3,204. In the 1870s Nottingham manufacturers were having disputes with the local trade unions and found they could employ non-union men from Long Eaton to work in their factories. Before the end of the decade the machines themselves were moved from Nottingham to Long Eaton and soon the large impressive mills were built on the banks of the canal which brought coal for their machinery. An even more rapid expansion of the population then occurred and by 1901 the figures had reached 13,045, the major industry being the making of lace. Fashions change and the demand for lace dropped markedly after World War I. Thus many of these fine buildings are not now used for the purpose for which they were originally intended.

At Sandiacre we reached the junction of the abandoned Derby

canal. The attractive toll house is now the headquarters of the canal preservation society and the area is one of great character and attraction. We moored for a few minutes to buy milk and provisions at Sandiacre itself. Here again efforts had been made to care for the canal, trees had been planted and a small grass lawn bordered the canal bank opposite to another fine lace mill. We cruised on under M1 and moored at Pasture lock for a cup of tea.

After tea we pressed on northwards through countryside dominated by the huge buildings of the Stanton and Staveley Ironworks. The old junction with the Nutbrook canal is now built over, but it is possible to note its position. We reached Stanton lock below which water is extracted from the canal for commercial purposes and it was here that we really met duck-weed in quantity for the first time. Great rafts of it lay below the lock and the lock itself looked more like a fine green lawn through which *Rose of Sharon* had to thrust to force a passage.

As we reached Barkers lock at Ilkeston, men on the road bridge called to us that we could go no farther. There was no water, they said, above the next lock—Stenson lock! It was already evening and we found a pleasant offside mooring beyond a pair of British Waterways maintenance boats and decided to leave our problems until next morning. We had covered 8½ miles on the Erewash since lunch, passing through eleven locks and we felt we deserved a break.

Next morning was misty, presaging a hot and sunny day. We were up early and as we finished breakfast, two British Water-ways men called across to ask us our draught. 'Eighteen inches,' we said, and this caused deep discussion. Apparently the pound above Stenson lock had had to be drained to allow certain emergency repairs and the water had been allowed to flow back from about five o'clock on the previous evening. 'It is still eigh-teen inches down,' they said, 'but if you can get under the rail-way bridge beyond the lock, you'll get through.' We let go at nine o'clock and cruised into Stenson lock.

Above the lock there was little doubt of the low water level.

The towpath wall was completely uncovered and reeds lay flat on the mud on the offside. The railway bridge was about a quarter of a mile ahead and the water was so clear that we could see the rubbish on the bottom. The two BW men were standing by to see if we came through successfully, but we ran firmly aground a few yards short of the bridge. The men then took a rope from us and pulled, Pearl pushed the bows out with a pole and I shafted from the stern. Inch by inch we were dragged through into deep water on the far side. 'You'll be all right now,' said our helpers and we were off again. I felt thankful for a $\frac{1}{4}$in steel bottom to the boat.

The next bridge was soon in sight and though we stuck again we came through by rocking the boat from side to side. So we proceeded until at last Shipley lock was in sight which, we felt, should mark the end of our troubles. We were in the centre of the channel a mere 100yd from the lock when we stuck again. Poling proved useless for the bottom was a black oily ooze into which the pole nearly disappeared. The bank was out of reach about 12ft away and it seemed that we should have to sit and wait for more water when a very helpful young man arrived and I threw him a windlass and asked him to open the paddle and let a lockful of water through to us. This he did, but we only drifted backwards a few yards and stuck again.

On the bank was a dreadful old man, gnome-like in appearance, who simply danced with glee and kept calling, 'You can't do anything, you're stuck, that's what you are, you're stuck!' I decided to climb on top of *Rose of Sharon*'s cabin and leap for the towpath. The gnome chuckled, 'You won't do it, you'll fall in,' and this prompted me to make an even greater leap which just carried me to the shore. The gnome then lost interest and commented that he might as well go for his pint of beer! I had to bring quite a lot of water through before the boat could move forward but she gradually came to the lock and pushed her nose in.

Even then she stuck again, this time on top of rubbish in the lock. We had now been joined by another helpful man and the three of us took the bow and stern rope and pulled the boat into the lock chamber. Once the gates were closed we passed through without any trouble. I had taken quite a lot of water through the lock and the pound above was now down a foot but we forged ahead slowly and steadily with no other difficulties.

Eastwood lock, the last before Langley Mill, is out in the country but not far from a council estate. I went ashore and started to prepare it when I heard an airgun shot. I carried on, but suddenly another shot sent a bullet past my hat. Very annoyed, I went to investigate and saw four youths running off. When I had opened the lock gates I set out through the bushes to see if they were still around and met them almost face to face, but they ran off again and that was the last I saw of them. In the final stretch to Langley Mill there was a certain amount of flannel weed but we reached the mooring below the last, derelict lock just before one o'clock.

We had our log-book signed at the junction cottage and picked up a postcard to send for our certificate, took a number of photographs of the old basin and the Nottingham and Cromford canals, had lunch and did some shopping and headed back south. By now there were a few more inches of water in the canal and we had fewer problems. Further, we had more time to look around us as we moved steadily down through the locks.

The valley of the Erewash is a natural line of communication for canals, railways and roads and we were interested to see the abandoned Nottingham canal running along the contour across the valley maintaining its level as we dropped down. Finally it turned eastwards through a gap in the hills and we lost sight of it.

At Gallows Inn lock we were helped through by a young couple who had seen us from their house and who looked forward to collecting their own newly built steel boat a few days later. We had noted Pasture lock as a possible night mooring and we asked them if this was a good place. They advised against it, though they gave no reason, and suggested the junction with the old Derby canal at Sandiacre. As we passed through Pasture lock we understood the cause for their advice for it lay beside marshalling yards where we learned that trucks would be shunted all night. We finally moored comfortably above Sandiacre lock for the night.

That night there was a terrible thunderstorm and I woke about midnight to see the whole sky lighted up. The lightning continued for a considerable time and the rolling thunder was muffled at first before it approached nearer to us but it was never really close enough to be frightening. Pearl and Lindy usually hate thunderstorms but they were both so tired that they hardly stirred. In the morning the weather had cleared and we had a perfect day for cruising.

My first task was to remove the weed-hatch and clear from the propeller an enormous ball of flannel weed which had made the engine lose power but had not brought it to a halt. It had been too large and firmly fixed to throw off by using the reverse. Then I found the battery connection had jerked loose and the self-starter would not work. This was soon remedied and our cruise through Long Eaton back to Trent lock was easy and uneventful. We crossed the Trent and moored in the beautiful waters of the lower Soar for a well-earned rest. Had the water level of the Erewash been normal we should have had no trouble at all. It was only lack of water in one of the best-fed canals in the country which turned our cruise into an adventure.

CHAPTER SIX

THE OLD GRAND UNION CANAL

SEASONED canal travellers often ask themselves which is the most beautiful waterway they have traversed. For us a dozen mental pictures spring to mind. The Leek branch of the Caldon canal in rhododendron time, the deep cuttings of the 'Shroppie' with the sun filtering through the tangled trees, the mighty Pontcysyllte aqueduct with the Dee tumbling below would all qualify. So, too, would the highland waters of the Macclesfield canal with the views across the Cheshire Plain, the sight of the Hatton locks from the third lock down, and Tixall Wide on the Staffs & Worcs canal with the birds coming in to the reeds to roost. We often vow that the loveliest waterway is the one we are cruising at the moment. Nevertheless there are some which seem even more beautiful than the rest and one of these for us is undoubtedly the top pound of the Old Grand Union canal. From the summit of Foxton to the Watford staircase are 20 lovely miles of quiet remote canal which winds almost as much as the Oxford through the rich hills of the Northants-Leicestershire border. When we traversed it in early August of 1969 we saw only five other moving boats upon it—three of these together in the long Crick tunnel!

We had cruised the Soar and had spent the night below Junction lock near where the old Wreak navigation, abandoned 100 years, had branched off to Melton Mowbray. We proposed to cruise through Leicester, up the many locks to Market Harborough and then back to Foxton to climb over the top and drop down into the Grand Union at Norton. We were up early and cast off at eight.

I have described the history of the navigations north and south of Leicester in *Water Rallies* and Hadfield has dealt much more thoroughly with this subject in *The Canals of the East Midlands*. Suffice it to say here that the first navigation was the Soar navigation or Loughborough canal permitted by Act of Parliament in 1766. The Leicester canal was permitted twenty-five years later, in 1791, and this terminated at West Bridge, Leicester, linking that city with the river Trent. We have an old print of West Bridge showing both narrowboats and sailing barges moored to wharfs below the bridge. In 1793 an Act was passed for a wide canal to link Leicester to Northampton on the Nene. The Grand Junction, whose Act was also dated 1793, was to be a wide canal and was to have a wide branch to Northampton, and Leicester was thus to be linked with London.

As so often happened in the canal building age money ran short when the canal had reached only as far as Gumley Debdale. It had climbed through twenty-four locks and a tunnel at Saddington, but was still far from Northampton. In 1805 an Act permitted the line to run to Market Harborough a few miles farther on and five years later the original line was abandoned altogether and another Act permitted the building of the Grand Union canal which was to link with the Grand Junction at Norton. This was built with narrow locks at both ends, though the two tunnels at Husbands Bosworth and Crick are of sufficient width for barges. This was opened in 1814 and at last Leicester was linked with London by water.

The link from the Grand Junction to Northampton was also built with narrow locks. The chief reason both here and on the Leicester Grand Union was the fact that they both link with the Grand Union between the great tunnels of Blisworth and Braunston. These tunnels had proved such bottle-necks to wide boats that such craft were prohibited after 1805 and only narrow-boats which could pass each other in the tunnels were permitted.

Thus there was little point in building the more expensive and more wasteful wide locks on the old Grand Union and south of Northampton.

Shortly after Junction lock we passed a small mooring for boats where two men were untying a hire cruiser and we cut down speed to pass. When we were directly opposite, the man at the bows fell in! One moment he was holding a rope with one hand and waving and smiling, the next he was dropping vertically into about 5ft of water. As he pulled himself out we tried tactfully to look the other way for there are few things more embarrassing than a ducking in front of strangers.

We continued southwards towards Leicester. It is not always certain in this section which is the lock cut and which the river, and at Birstall we found the lock rising alongside us as we cruised by and had to reverse about 100yd and start again. Some children, obviously enjoying the scene, told us that this was a common occurrence!

At last we reached Belgrave lock, cruised along the stretch where the narrowboats had been moored during the National Rally in Leicester in 1967, and continued to West Bridge. This was lined with interested spectators and we recalled how often we had looked over the bridge hoping to see a boat. Pearl was born in Leicester and it was a great thrill to us both to cruise through the heart of the city. From now on it was all new ground for us.

The canal in Leicester is wide and deep and it seemed to us such a pity that the city makes so little of it for it is well constructed and could be made beautiful. The short stretch by the castle has public gardens running almost to the water's edge but even these seem to turn their backs on it. The length south to Aylestone, passing the electricity power station, the gasworks and numerous factories and waste ground, was typical of so many of the less attractive urban canals. We passed through Freeman's Meadow lock and wondered how long it was since Freeman tended his meadow there. Incidentally this lock was filled by one of the largest sluices and heaviest paddles I have ever seen.

It could all look so attractive with a little effort and not a great expenditure of money. We continued to Kings lock now out in the country and said good-bye to the river Soar.

The twenty-four locks to the Market Harborough pound made very heavy going for all are wide and some had only one paddle working. Most were against us and some of the intervening pounds were low and weedy as we climbed slowly all through a long hot afternoon. We passed one moored boat *Reward* at Wigston and the people aboard called out, 'There's *Rose of Sharon*!' At the next bridge I went back along the road to a shop to buy some milk and found the boat owners in residence. They were shutting the shop for a fortnight and were going on their boat to Stratford and they told me they had fitted out the boat entirely themselves. They had read my earlier books, particularly the chapters describing the fitting out of our two boats and had recognised us as we passed. Two days later we saw them going well on the top found.

At six o'clock Pearl said, 'Let's have a rest and go on after supper,' so we moored, not without difficulty, for the water by the bank was shallow, and enjoyed some delicious pork chops. Like giants refreshed we carried on for a further two hours, finally mooring at nine o'clock above Newton top lock. We had cruised for eleven hours out of thirteen on a very hot day and had covered 19 miles and twenty-five heavy locks and were not sorry to tumble into our bunks. We could not help chuckling at a notice above the last lock we had just passed which stated that these locks would be closed from 7pm until 8am!

Next day was Saturday and we were up early and let go again at eight o'clock for we still had five more locks to pass before reaching the Harborough pound. They were all for us, but there was very little water in the intervening pounds. Above the second *Rose of Sharon* went firmly aground and I had to bring water down which could ill be spared from the pound above. However, we came through at last after nearly two hours and set off along the deep and beautiful Harborough pound. We took eight minutes along the $\frac{1}{2}$ mile Saddington tunnel and an hour later reached the basin below the Foxton locks. Here there was great activity for the Junction Cut boats were preparing to start with the large parties which they take for a week's cruise.

We were making for Market Harborough, however, and turned left into the winding Harborough arm. We had arranged to meet Pearl's brother and his wife at the swing bridge beyond the junction and they were waiting for us with baskets of provisions including a delicious cooked chicken!

The 5 miles to Market Harborough form one of the finest examples of a contour canal that we have ever cruised. The actual distance by road is about half that by water and the final mile crosses the main road, winds round a low bluff, recrosses the road again to the south to double back finally to the wharf only a few hundred yards from the first road crossing. Much of the length is well wooded and forms most attractive cruising.

We moored for lunch before we reached the wharf as we knew there would be great activity and sure enough a stream of boats came out during the afternoon. When things became quieter we moved in and paid our respects to Giles and Frank Baker who run an admirable boatbuilding and boat-hiring business and have done much to support the canals throughout the country. Poor Giles had his leg in plaster and told us he had broken it lock-wheeling in the middle of the night, taking a boat through to one of their other bases at Trevor on the Llangollen canal. He hoped to be free of his plaster the next day.

In the morning we returned to the pool below Foxton locks. Much has been written about this wonderful flight of locks and about the remarkable inclined plane which lies alongside. We had visited them some years previously by car but this was the first time we had come by boat and our immediate impression was of their beautiful condition. The locks themselves were well maintained, many of the gates seemed new, the surrounds were mown and the towpath tidy. But the great difference since the earlier visit was in the side ponds which were dredged and in perfect order. When we had last seen them they were completely covered with weeds.

The principle of the locks is simple. The ten locks are grouped in two staircases of five with a small intervening pound to allow boats to pass in the middle. Each staircase of five locks has five pairs of gates and a top gate, so that the top gates of the lowermost lock form the bottom gates of the lock immediately above. Alongside the locks are the side ponds whose levels are equal to the levels of two adjoining locks when the intervening gates are open.

When we arrived, a boat was just about to ascend, so we waited for it to pass through the bottom lock, which we then emptied and entered. We then emptied the lock above into the side pond and, at the same time, filled the bottom lock from that side pond. It was all so simple as step by step we climbed and two other boats followed us. In the pound between the two flights, a descending boat was waiting. He had quite a long wait with a procession of four, but the scenery is so lovely that I doubt if he minded. At last we reached the top after an hour's climb. I made one mistake which I regretted for I had suggested to Pearl that she should nose the upper gates in each lock, but had not put up our front cover. All was well most of the way, but we reached a pair of leaky gates and water poured into our front cockpit and even over the step into the front cabin. All would have been dry with a cratch and our front cover would have protected us equally well. The cabin carpet took some time to dry out.

At the top of the locks I went to examine what was left of the inclined plane. The story of the plane has been told by many authors, notably Rolt (*Inland Waterways*) and Hadfield (*The Canals of the East Midlands*). Built as late as 1900, it was a wide concrete slope with rails running from top to bottom on which were caissons on wheels, each capable of carrying two narrow-boats. Unlike so many of the tub-boat inclined planes of Shropshire, the narrowboats were carried up sideways. The plane was so wide that two caissons, each carrying boats could pass one

F

another, the one ascending and the other descending. There was a short arm of the canal from above the top lock and a basin at the bottom and the boats could be floated in and out of the caissons. The motive power was a steam-engine and this led to the failure of the whole project despite the fact that it was a most ingenious piece of engineering. For it to be of value to boats, the engine had to be kept in steam and few boatmen were prepared to wait a couple of hours for steam to be raised. Traffic was not sufficient to keep the engine permanently in steam and ready for action. Within a very few years, the plane was in use during certain hours of the day only and the locks were used at night, and later the plane was abandoned altogether and the locks came back into their own.

In many respects the failure of the plane was a pity for it saved both time and water. Unfortunately, it came a century too late when traffic on the canal had already dwindled. It was also the last attempt to make the canal suitable for wide boats for each caisson could have carried a barge. Had it been successful, a comparable plane or broadening of the locks would have been necessary at the south end of the canal at Watford and this was under consideration. The tunnels had been built wide enough but the locks still restricted the boats to 7ft beam.

All that was left of the plane was submerged in a deep jungle of thorn and bramble. It had rained for most of the morning and was still raining quite heavily as I forced my way through waist-high prickly undergrowth beneath a cover of saplings. Had I not seen pictures of the plane in action, I should not have made out much of its construction. I was not sorry to rejoin *Rose of Sharon* for lunch and peel off my soaking garments. After lunch the rain stopped and we set off along the beautiful and remote 20 mile top pound.

It is strange how little is written of our English South Midlands. There is a Jurassic limestone and marlstone belt (Jurassic is the name of a geological system of rocks first described in the Jura mountains) which underlies much of the county of Dorset. Passing through Somerset and Gloucestershire, the strata form the bold scarps of the Cotswolds. Edgehill in Worcestershire is part of their continuation and throughout these regions the countryside and the villages are famous for their beauty. The same rocks are also to be found in Yorkshire, on the coast round Scarborough and Whitby and forming the scarps of the Cleveland hills. In Lincolnshire they are much thinner but they frame the splendid ridge on which Lincoln cathedral stands so boldly. Between Lincoln and Edgehill the rocks are also present and still build scenery of great beauty. Ironstones are amongst the limestones and the villages of South Lincolnshire, East Leicestershire and Northampshire are tinged a warm brown colour from that attractive stone. The countryside of the shires is famed for hunting but is little known to the ordinary tourist and through this rich, rolling highland, the canal runs for the next 20 miles. We set off and were delighted to find how deep and clear was the channel.

Our aim for the night was Welford and we were keen to cruise up the little Welford arm. We first passed through Husbands Bosworth tunnel, the second of the three on the canal and on by the little mooring within reach of the village. Continuing, we came to a fine wide junction and turned left towards Welford where, not long ago the little arm was semi-derelict and clogged with weed. Now British Waterways have cleared and dredged it and it is in excellent condition. It is only 1½ miles long and has one little lock, now completely restored, with a rise of less than 5ft. The canal itself is really a navigable feeder, carrying water from the large reservoir at Naseby. Beyond the lock is a winding hole and the last 300 yards lead to the terminus along the north side of which is an interesting and attractive range of canal buildings which have clearly seen better days. The canal is narrow and the terminus ends abruptly, but it seemed certain that at one time the basins were more extensive. The last building in the range is an inn which we had heard served excellent meals, but this proved to be a quiet evening and we had given no warning of our coming. We reversed to the winding hole and turned, mooring for the night half-way to the junction in a quiet and peaceful rural berth. After supper, a beautiful sunset illuminated the wide sky and the rich rolling fields.

Back at the junction next morning we continued south. There is little to say of the canal except that it winds pleasantly through charming country with wooded hills to the east and glimpses of valleys, at least one water-filled, to the west. The waterway makes an almost complete circuit round a little hill before reaching Yelvertoft, reminiscent of the Oxford canal at its most wayward. We moored and walked into that village though Pearl had a long way to go to find a dairy.

Farther south, we came to the last and longest of the three tunnels, that at Crick. The whole waterway had been so quiet that we had seen only two other moving boats, but here we met three, all close together in the tunnel. Though there is room within to pass, we saw the first before we entered and decided to wait. As we watched, we could see the bobbing lights of the other two and when all three had emerged, we went through.

Eventually we came to the flight of the Watford locks at the south end of the long top pound. There are seven in all, the first single and the next four in a staircase similar to those at Foxton. We had often noticed the beautiful painted gates and foot-bridges from the London train. The line north from Euston runs beside the M1 motorway for some miles. Beyond the 'Services' there is a short cutting and beyond this the locks can be seen stepping sharply up the hillside. Now we were looking down on the great main line with its bustling trains.

The staircase has side ponds similar to those at Foxton and paddles connecting the locks to them. A notice-board with a diagram showed exactly how they should be worked. When we returned to this canal in 1971 the system had been simplified still further and the tops of one set of paddles were painted red, and the other set white. Thus the instructions could tell when to raise the red and when the white paddles.

Below the locks we crossed beneath the motorway and the main railway line and cruised out into the country again. We were looking for a suitable night mooring away from the roar of cars and trains and had difficulty in finding the right place as the canal sides are very shallow. Eventually we were lucky and tucked ourselves in.

A few hundred yards away a hire boat had preceeded us to a mooring and its occupants walked along to have a chat. They had come from Market Harborough and were enjoying their holiday, though they commented on the weediness of the top pound, a point on which we disagreed. They told us that they had had to remove the weed hatch on several occasions to clear the propeller and then came out with a most alarming statement: 'We finally decided to leave the weed hatch off altogether!' I was perhaps a little tired and the full implications of this did not strike me at the time. But then my mind boggled. Perhaps the turning propeller would throw up only a little water through the uncovered hatch into the boat if it were going forward, but in reverse it must surely flood the boat completely. Now I knew how to sink the unsinkable! I recalled the manager of a hire firm telling me that one of his 'unsinkable' boats had been reported sunk on the Shropshire Union canal by the party who had hired it only hours before. Perhaps here the weed hatch had been removed or put on loosely and replaced later, under water by an embarrassed hirer. It might be a good safety precaution to print on the cover 'Do not remove while the engine is running'. In any case, it would prevent 'little Willie' from engaging gear while father was groping round the propeller.

Next morning, we let go and cruised through the swing bridge at Norton junction and out into the main Grand Union canal near Buckby Top. It had been a lovely cruise along one of the most remote canals with everything that we have ever sought on our journeys. It had beautiful scenery, interesting engineering and, above all, peace.

CHAPTER SEVEN

THE GRAND UNION CANAL
AND ITS BRANCHES

THE Grand Union from London to Birmingham is the major canal in the whole network, the most expensive to build and it retained its commercial traffic longer than most others. The main line was formed by an amalgamation of quite separate canals built in the last decade of the eighteenth century. The most important section was the Grand Junction from the tidal Thames at Brentford to the Oxford canal at Braunston, a distance little short of 100 miles with two summit levels. This received its first Act of Parliament on 30 April 1793 and was estimated to cost half a million pounds. In the event, with problems over its two great tunnels of Braunston and Blisworth and its aqueduct at Wolverton, the total cost was over a million and a half.

From the Birmingham end there were two separate canal companies, the Warwick and Birmingham, whose Act was obtained on 6 March 1793, and the Warwick and Napton whose Act was twelve months later. For short canals with no major structures, these two were fairly expensive for they were heavily locked, dropping down to the Avon valley from the higher ground of Birmingham and Napton. They were completed and opened in 1800 and provided a through route to the Oxford canal at Napton and thence by the level 5 miles to Braunston and the Grand Junction. This was opened throughout in 1800, except for the Blisworth tunnel which had a plateway running over the hill until 1805 and traffic from the Midlands to the Metropolis had no more need to use the Thames, in those days a poor navigation.

The Grand Junction to Braunston was built as a barge canal with locks over 14ft wide. The other two, perhaps due to the heavy locking, had 7ft locks though their bridges and the Shrewley tunnel were built wide enough for barges. Barge traffic was very soon prohibited on the Grand Junction owing to the bottle-necks of the two great tunnels which had to be legged through with men lying on their sides on the boat and walking their feet along the sides of the tunnel. In the 1930s, when unemployment in the country was at a high level, the narrow locks from Napton to Birmingham were replaced by wide ones and barges of 12ft 6in beam could then make the whole journey. This was not the hoped for success, for though a powered barge could run quickly through the tunnels with comparatively little delay to those coming the other way, two such barges deeply laden had difficulty passing each other over much of the canal, each needing the central channel. However, it was of great benefit to narrowboats running in pairs which could share the locks and pass through much more quickly.

We have not cruised the whole of the Grand Union at any one time but have covered most of it in sections. We have entered at Lapworth to leave at Norton and have cruised in at Braunston to turn off at Southall where the Paddington branch forks eastwards to join the Regent's canal and London docks. After our trip through the Stourbridge and Dudley canals, we decided to make for Stoke Bruerne and entered the Grand Union at Lapworth from the Stratford canal. This route from Birmingham via the Worcester Bar and Kingswood is less heavily locked than the alternative way by Farmers Bridge. At Lapworth the wide, deep canal is already in rural surroundings and we cruised south-westwards towards Warwick. It was a hot sunny day as we approached the short Shrewley tunnel and saw the towpath running up the hillside. A cruiser was inside and we waited for it to emerge. When it came out the steerer, stripped to the waist, called for a towel and commented as we passed him, 'You'll need an umbrella—it's raining in there!' It turned out to be one of the wettest tunnels I have cruised.

We had planned to reach the famous Hatton locks in the late afternoon and to moor above and photograph them before nightfall. We would then lock down on the following morning. In the event, we were early and reached the top lock sooner than we had intended. As we approached there was great activity on the lock which was being emptied and I went ashore with my camera. Suddenly the people on the lock saw *Rose of Sharon*, stopped emptying the water, raised the top paddles and heaved open the gates. Before Pearl realised what was happening, she had steered into the lock and we were in the flight! There are twenty-one locks with short intervening pounds and boats are not allowed to moor for the night in the flight. Thus, with nineteen Lapworth locks behind us already, we had a further twenty-one to do before we could finish our day's cruise.

The Hatton locks have large enclosed paddles and huge sluices. The locks fill and empty quickly but each paddle requires twenty-three turns of the windlass to open it fully. As there was a boat some way ahead of us and most of the locks were empty, this meant ninety-two turns for each lock and the task seemed formidable.

Below the third lock are the workshops of British Waterways and the canal curves slightly but from there the locks extend in a straight line down the hillside and there is a fine view over the Avon valley with Warwick in the distance. We continued down, lock after lock, while the hot afternoon merged into a warm evening. As we neared the bottom we kept asking bystanders on the towpath how many locks were still ahead and the answer always seemed to be 'four'. At last we reached the bottom and turned left, while the line of the old Birmingham and Warwick canal continued ahead to the now abandoned basin. It seemed a pity that the beautiful city of Warwick should have isolated itself from its attractive canal. It also seemed a pity that the earlier narrow locks of the Hatton flight were not still in use as well as the wide ones as this would have saved so much water. Most of the chambers are still visible but have been partially filled in and used for overflows.

Next day we went through Leamington on the least attractive stretch of the whole navigation. This proved to be a good shopping centre and a convenient place to meet passengers and my sister and her friend joined us for a few days, finally departing at Coventry.

The locks up to Napton are as large and numerous as those at Hatton but come in a series of shorter flights. We eventually turned left into the Oxford canal and saw the Napton windmill standing starkly on the hill ahead of us. Five miles brought us back on to the Grand Union at Braunston, surely the most exciting place on the whole canal system. It is an important canal centre and a junction which must have been full of commercial activity in bygone days. On that occasion Michael Streat still had his small fleet of narrowboats carrying coal from Atherstone to Watford as well as his huge marina, his workshops and dry docks and his hire fleet. Now he has joined with Ladyline but, alas! the coal trade has finished. This must be due at least in part to the shallowness of the Coventry canal between Nuneaton and Atherstone caused by mud washed in from the stone crushings on the canal-side. In 1969 we were moored on that stretch when Mr and Mrs Collins came by in *Stanton* and *Belmont* carrying a load of coal and such was the channel that it had taken them three hours to cover 2 miles. It is beyond my belief that such pollution with stone dust can be tolerated in an otherwise well-constructed and beautiful navigation.

At the base of the Braunston locks are more workshops where the Willow Wren narrowboats operated for so long. From here the six locks climb up to the first summit level and pierce the Northamptonshire ironstone in the 2,000yd-long Braunston tunnel. This tunnel is peculiar in that it has a slight dog-leg near one end. The story goes that the direction was not quite right and that a dog-leg had to be introduced to correct it.

Though the tunnel is high and wide enough to make it possible to see from end to end, these bends block the view. When we followed two other boats through from the south on one occasion, there were several boats coming the other way but the tunnel is sufficiently wide to pass other narrow-beam boats and we had no difficulty with the first two. The third seemed to be swinging from side to side as we crept slowly along the wall and as it drew level it gave us a fairly hefty sideways nudge. It was a much lighter boat than ours and the bump must have seemed shattering—we couldn't hear what the steerer said but he blew his horn loud and long to relieve his feelings! When we caught up the other two boats at the locks they told us that it had also bumped them both in passing. As there were yet other boats behind us the tunnel trip must have been very uncomfortable to the crew.

A couple of miles farther on is Norton junction, where the Leicester section of the Grand Union comes in from the north, and beyond it is Buckby Top lock. The flight of seven Buckby locks are well separated and cover over a mile of canal. In summer they are very busy and it is here that a lock-wheeler on his cycle would be of great value. Unless sharing locks with a second boat I have so often seen the lock ahead taken and emptied for a boat coming up when a moment's pause would have let *Rose of Sharon* go down in it. On this occasion my sister opened gates while I walked ahead to see that no one stole a full lock from us.

Below Buckby the canal runs between the M1 motorway and the main line railway from Euston to the north. On the one the trains flash by at more than 80 miles an hour and on the other there is the continuous roar of fast-moving motor traffic. Both are shielded by trees over much of the route and on the canal progress is slow and peace reigns. This is a long level pound and at last Gayton is reached with its little branch down to the river Nene. Ahead is another ridge of hills with the mighty Blisworth tunnel running beneath them.

Blisworth tunnel was the most difficult and expensive obstacle on the whole navigation. Though the rest of the canal was open by 1800, it was five more years before the tunnel was completed and during those five years a plateway with waggons was laid over the hilltop. At 3,075yd (plus or minus a few, depending on the authority) it is the longest still navigable tunnel on the whole canal network.

Coming up from the south on a later occasion we met a boat in the middle, apparently stuck. As we crept ahead, it turned out to be a steel hull with an outboard engine attached, or rather, the outboard which should have been attached had fallen off into the water. The steerer had retrieved it and was hoping to get it to restart so we offered a tow which he politely declined as he was going the other way.

A few hundred yards beyond the tunnel is Stoke Bruerne, a delightful canal village astride the paired top locks. The warehouse is now the Waterways Museum with its maps and plans, cans painted with roses and castles, lacey plates, a restored cabin and a wealth of attractive canalania. It is a very good museum, attractively displayed and well labelled, and we have spent several happy hours in it. The adjoining houses are of the same date as the canal and the waterside pub opposite is equally attractive. The whole village nestling in the hills and busy with boats is well worth a visit.

On our second passage through Stoke Bruerne, we were going right through to London. Below the Stoke locks is a long pound through the village of Cosgrove which has a little pedestrian tunnel from the towpath side to the shops. The single lock at Cosgrove is at the junction with the now abandoned Buckingham arm and below this the canal runs on a long embankment over the valley, crossing the river Great Ouse on the cast-iron Wolverton aqueduct. At one time locks dropped into the valley and climbed up the other side, wasting a great deal of water. When the embankment was built, it ran into trouble for part of it was washed away and then the three-arched aqueduct collapsed. The present structure was built in 1811 and has stood the test of time. Though only 40ft above the river as compared to 121ft on the Pontcysyllte near Llangollen, looking over the edge from a boat gives a great impression of height.

From Wolverton onwards the long line of the Chiltern Hills comes into view and the canal winds slowly towards them stepping higher a few locks at a time. The final climb to the summit is at Marsworth, another attractive village with a pub which serves tasty snack lunches, and away to the right are the reservoirs which feed the top pound of the canal. At Bulbourne, above the top lock, the canal enters a long wooded cutting which hides the little town of Tring on the right. We cruised through this on a very hot day and were grateful for the shade as we moored for tea. At the other end of the pound is the famous Cow Roast inn which gives its name to the top lock of the long, long flight which steps gradually down to London.

We visited the Cow Roast inn on our return journey from London and though no cows were being roasted, we had some excellent smoked salmon sandwiches. We inquired about the origin of the name, spelt Cow Roost in earlier accounts and learnt that it originated from cow rest. The ancient drovers' road used to follow the gap in the hills and cattle were driven this way into the valleys in the winter and on to the hills for grazing in summer.

It was noticeable that there were fewer pleasure-boats in the heavily locked stretches than in the gentler cruising grounds beyond the Chilterns. This was no doubt partly because of the locks themselves and partly because the canal passes through so many towns in the Colne and Gade valleys. Despite the urban surroundings the navigation is often very beautiful, running in and out of the rivers much of the way. We saw many factories and wharfs which had once taken goods by water but only the Croxley Paper Mill at Watford still received coal by this means. Even this trade was to stop in 1970.

We moored for the night below Common Moor lock in sight of the coal boats on a little island between the canal and the river Gade. I am afraid the obvious pun occurred to us but the site itself was equally inviting. Next day we completed our locking at Cowley Peachey and entered the long level pound which extends the whole way to London. The Slough arm turns off to the west at this point and a few miles farther is Bulls Bridge, Southall and the branch to Paddington. On that cruise we were making for the rivers Lee and Stort but that story is told in another chapter.

The Grand Junction canal in its day had many branches. In addition to the Paddington branch and the Slough arm just mentioned, there was a short length at Watford and a little arm through a single lock to a group of canal workshops at Rickmansworth. At Bulbourne on the summit a navigable feeder came in from Wendover and below the Marsworth locks a branch stepped down to Aylesbury. There was a short canal at Newport Pagnall and an 11 mile length from Cosgrove to the ancient town of Buckingham. At Gayton, north of the Blisworth tunnel, a narrow canal was constructed to link with the river Nene and at Norton by the Braunston tunnel the Old Grand Union was built to cross the hills to Foxton and Leicester. Apart from this last important main line, there were nearly 50 miles of 'collaterals and cuts', half of which have now been abandoned. We determined to explore as much as possible of these lengths by boat if we could and by car and foot if *Rose of Sharon* could not go.

When we reached Cowley Peachey we had a day in hand as we proposed to avoid the weekend in our journey through London. This gave us the opportunity to turn into the Slough arm and see something of the work of the London and Home Counties branch of the IWA and the Grand Union Canal Society who have been striving to save it from abandonment. They have brought out an attractive booklet entitled *Slough's canal—the future?* showing how its development would benefit the whole area round Slough and their volunteers have removed rubbish and have generally maintained both the channel and the towpath. On a grey Saturday afternoon, having shopped at Uxbridge, we locked down through Cowley lock and turned into the Slough arm just as the rain started to fall. Against the bank by the junction a number of maintenance boats were moored and the bank itself was dumped high with rubbish. There were bedsteads, bicycles, cisterns, lawn-mowers, carpets and every conceivable item for which people have no further use. We wondered if this was the general dump for the south end of the Grand Union canal or whether it had all come out of the Slough arm. We knew the working parties had done wonders but this seemed to be a vast pile of rubbish. We also wondered how much more there was still in the cut and this gave us plenty of food for thought as we turned in.

At first sight the Slough arm is seen to be a remarkable canal. The map shows it to be practically straight from one end to the other with only two slight bends in its 5 miles and this directness is its most striking feature. It first crosses the channels of the river Colne on a number of small aqueducts from which we could see the gravel pits and the wide area of water. We were not entirely surprised to read later that there is a greater acreage of water in this valley than in the whole of the Norfolk Broads. The fact that their levels are different would make it difficult to link up the flooded gravel pits and reservoirs into a single system. Such a scheme has been considered by the Colne Valley park authority with the canal as one of the more important links, but the maintaining of water levels poses a difficult problem.

West of the valley we entered a long wooded cutting and near one of the bridge-holes we met a couple of cruisers starting out in the rain from the Iver Boatyard. We were their first hazard to meet and pass and their steerers gazed at us with anxious faces as they came by. This section of canal would have been beautiful in sunshine but was not at its best in the dripping rain. Fortunately the sun was to appear on our return the following day but now we plodded slowly through the very wet cutting. The waters of the canal were crystal clear and though the sides were weedy the central channel was deep and free of weed. Large areas of white water lilies grew in many places in great profusion. This is a more attractive flower than the yellow variety whose leaves are sucked under the water as the boat glides by.

Beyond the cutting we came upon the boatyard but as it was a busy Saturday when the fleet were turned round we thought it would be hardly fair to stop. There were many private boats moored and as we nosed through we asked a young man working on a narrowboat if we should have any difficulty getting through to Slough. 'I've never been,' he said, 'though this boat once made the journey.' This seemed odd for Slough was only 3 miles away! 'If you get fed up', he called after us, 'there's a winding hole in about another ½ mile and you can turn back.'

A few hundred yards farther on, a very long boat was trying unsuccessfully to turn in the width of the canal. Both bow and stern were stuck and the steerer was shafting first from one end and then from the other. The boat straightened up as we approached and we were able to follow it. At this point we were in a length bounded on the north by extensive gravel pits and on the south by an overgrown towpath, a high fence and the back of a factory. It was five o'clock and still raining heavily, and I felt that we had had enough for the day so we ran our gangplank ashore, not without difficulty, moored and put up covers for the night and then stripped off our wet things. Though close to large centres of population we were remote and almost inaccessible. Pearl cooked a tasty meal and after it we settled down for the evening to watch a good programme on our little battery television. The whole occasion stands out in my memory as being perfectly delightful and most unexpected. The wet night, the less well-maintained canal, the partly industrial area, yet our snug boat isolated us from everything and gave us peace and contentment.

Next morning we woke early to a brilliant sunny day, breakfasted, and set off again for Slough. Soon we were amongst factories and then anglers lined the banks though the weed was worse and the clear channel little wider than our boat. We were welcomed as we cruised by for we were helping to keep open the channel.

The weed extended right across the canal for the last ½ mile and no one could fish except in one or two clear pools. *Rose of Sharon*'s big slow propeller chopped steadily through and occasionally I engaged reverse to throw off all that had collected round the shaft. The water was still clear and fish darted about, but there was no room between the weeds for the hook and line. The last ½ mile was very slow but we kept moving towards a goal which we had already seen for some time. We were now in a pleasant residential area with houses on one side and a park on the other and a wide basin at the end abutting on to a main road. The basin was fenced all round, with a timber yard and a coach park on either side of it. We moored at about ten-thirty and Pearl went boldly through the coach park though I was rather afraid it might have been patrolled by a fierce dog. In due course she returned unscathed carrying a few necessary provisions.

It seemed such a pity that the timber yard did not still receive its timber by boat and that the pretty little basin was not thronged with busy people. We read of the proposed marina which would be ideally sited here but we also heard of the suggested plans for abandoning the last mile and using the reclaimed land for a road. If this happens there will come a time when Slough will have cause to regret that it pushed its canal farther from the centre. Feeling strongly about this, I wrote to the mayor on my return home and received a most courteous reply. The work of the volunteers deserves the fullest support for this little canal could be an asset to Slough and Slough could be a welcome port of call for the pleasure boatmen. In addition, we had seen enough to know that the anglers wanted the water if boats would keep it clear of weed.

Viscount St Davids, that far-sighted and far-travelled boatman and organiser of the Regent's Boat Club, has suggested a possible link with the Thames at Slough. In an article in *Motor Boat and Yachting* (December 1964), he showed how a 2 mile cut could be constructed to join the canal to the Thames at its nearest point. Such a link would be of great value to both

Thames and canal boatmen who would not then need to enter the tidal Thames at Brentford to make the journey between the two. A time will certainly come when the numbers using the waterways make the building of new canals imperative. It is therefore never too early to get the schemes on paper so that redevelopment of the area will not jeopardise their fruition.

The history of this little waterway is unusual in that it is one of the last canals to have been built in the country. Constructed between 1879 and 1883, it appears from the map to form half of what might have been a short cut from the Grand Union to Maidenhead. Throughout much of the nineteenth century the condition of the Thames as a navigation was poor with too many flash locks—weirs with gates in them—and too much weed. The building of canals to avoid the upper reaches was largely due to this fact and the Wilts & Berks and North Wilts linked Abingdon with both Kennet & Avon and Thames & Severn. Had the Maidenhead-Cowley link been built, London's connections with the West Country would have included only short lengths of the river. Hadfield in *The Canals of South and South East England* describes the many attempts to obtain the necessary Act of Parliament. All failed, and the need for such a link in the second half of the nineteenth century receded with the improvement of the river and the coming of the railways. Today the circumstances are changing again and the need for such a link is becoming more apparent.

After half an hour's stay and a chat with some people who lived in one of the houses adjoining the canal, we turned and set off back towards the junction. Once more we had to plough our way through the weedy surface until we reached the clearer water. We passed the boatyard, entered the long cutting, recrossed the Colne valley and finally rejoined the Grand Union. We wished we could have made similar visits to Buckingham and Wendover.

The Buckingham arm was still legally navigable throughout until 1961, though as early as 1904 de Salis noted that the last mile above Maid Morton Mill was extremely shallow. The 1¼

G

miles to Old Stratford was considered by the Ministry of Transport as a possible length for moorings though the rest was beyond redemption. The history of the canal and its state in 1962 was recorded by E. E. Kirby in *Country Life* of that year. Even a few months only after its legal abandonment, the canal was very far gone. The terminal basin had shrunk to a small weed-covered pond and the towpaths were completely overgrown.

Kirby described how the short wide section to Old Stratford was included by the main Grand Junction Act of 1793 and the 10 mile section from there to Buckingham, built to the narrower gauge, received its Act the following year. The intention had originally been to lock down to the Ouse at Old Stratford, but the extra locks would have increased the cost and the plans would have met opposition from the millers. The Buckingham section was completed in nine months only—a record time made possible by the ease of the terrain. He also described how steam launches were built at Stony Stratford a mile from the canal, dragged by traction engine to the wharf at Old Stratford and launched sideways into the canal to be sent by water to customers as far afield as the river Nile!

We had passed the junction at Cosgrove on several occasions and had wondered whether to push *Rose of Sharon*'s nose as far as possible up the Buckingham arm. The branches hang far out but a start would not be difficult. However, we decided to see what we could of the canal by car first to various points and thence on foot. We made for Old Stratford wharf where the canal divided, a short arm running up to the road and another continuing westwards. Everything was deeply overgrown and there was not a drop of water in either section. It would not take a lot of cleaning up and dredging and there is room for a marina which would be handy for the old A5, Watling Street trunk road. It looked very dead but could easily come alive again.

We continued up the main road towards Buckingham. For a canal which was abandoned so recently, we were staggered at its obliteration. Many rural canals, abandoned much earlier, are still in water and are apparently almost fully navigable. It seemed as though people had hardly waited to fill it in and have done with it. It swings right round the small village of Deanshanger and here road bridges have been dropped and the tree-grown bed is hardly visible.

Westwards of this village the canal is never far from the young river Great Ouse starting its journey to the Fens. The valley is wide and open, quiet and rural. We were keen to see what was left of Hyde Lane lock, one of the two little locks which kept the canal a few feet above the valley floor. A grassy path skirts some attractive lakes and crosses the canal by a small wooden bridge, no doubt once movable. The canal itself is very straight but quite waterless and it was even difficult to decide which side had been the towpath. Perhaps there is an abundance of fishing in the river Ouse and in the lakes but I could not help thinking that water, if possible moved by the occasional emptying of the lock, would have attracted many anglers. It would certainly have attracted *Rose of Sharon*. We walked westwards towards the lock which we found surrounded, in fact overgrown, with trees. Below it was an overflow and what little water there was trickled through to the river. The brick chamber was not too decayed and the rise was only about 6ft. There were remains of rotted gates.

Our next point was Maid Morton Mill, approached by a little lane from the main road. The mill is now a very attractive dwelling house and the canal used to run along its garden wall. I say 'used to run' for now it is completely obliterated. Bulldozers have moved in and covered over the channel whose course is marked only by two lines of trees which used to form part of a hedge. This showed how easy it is to eliminate a rural canal running by a river for it performed no useful task of drainage or carriage of water. No doubt the removal of the barrier and the increased access for farmers was an advantage, but it seemed to us a great pity for we can ill afford to lose 11 miles of quiet water at a time when this country needs every yard of countryside amenity.

We then sought, but did not find, the second little lock on the waterway. We walked through some allotments to the line of the old canal but its use today is a tip for rubbish.

In Buckingham we took petrol at what had once been the town basin but is now a garage with a wide open space. The wharf-house still stands and one or two other buildings which show something of the size of the basin before the bulldozers moved in. We wondered if anyone in the pleasant little town of Buckingham regretted the loss of what could have been a beautiful marina. It is central, a few paces only from the shops, a wonderful little port of call for visitors in boats and if I had lived in Buckingham I know that this is where I should like to have moored *Rose of Sharon*. It seemed to us that it was the little canal that no one wanted. It was clearly very far gone in 1961 when Parliament pronounced its death knell. We wondered what the local people think about it now—do they prefer the empty space to the wharf, the tipped rubbish to the quiet water, the bulldozed length to the flower-filled towpath? Do fishermen regret the passing of both water and fish? We could only record our own regret that once we could have cruised to Buckingham and now it was too late.

The Aylesbury arm is a very different matter. It descends from Marsworth Junction through a flight of well-kept narrow locks, each with only a 6ft fall, to the splendid basin in the busy town of Aylesbury where many boats are moored and there is great activity. The locks from the main line come thick and fast and are then gradually spaced out. The rounded masses of the Chiltern Hills are always on the skyline to the south.

It was on this little canal that the Inland Waterways Association held the National Rally of boats in 1961. Mr Meinertzhagen had started a boatyard there two years before but the canal was still in danger of abandonment. It was said that the local planners had their eyes on the basin for the usual reason—a car park! (When it is not a car park, it is a bus station.) The visit of 100 boats, the gay scene and the publicity attached to it proved the turning-point and the canal was saved.

Unlike the Buckingham canal, it seemed to us that the Wendover arm, sanctioned in 1794 in the same Act as that which permitted both the Buckingham and Aylesbury arms, and opened to traffic as a navigable feeder in 1799, could (and should) be reopened tomorrow. Bulbourne Junction at the top of the Marsworth locks leads to a canal open and used. We made for Tring wharf and the bridge over the clear water once more showed a canal in good condition. We paused next at a small road bridge at Little Tring.

Here the canal was waterless and the towpath ran eastwards past the gardens of an attractive house before diving into undergrowth. After about 100yd we came upon six narrowboats moored but floating! At the narrows beyond the end pair the canal was filled in but from here eastwards there was plenty of water which bubbled up through a pump-house from the great feeder reservoir by the Marsworth locks. Here, then, was the reason for the excellent condition of the first mile of canal, it was still an important feeder for the top pound of the Grand Union.

The canal lies close beneath the Chiltern scarp and winds amongst its outliers, its whole course on porous chalk. In consequence there may be problems of preventing leaks, though other canals, notably the main Grand Union top pound, also rest on a chalk substratum. This rock is porous but does not have the wide open joints of such limestones as the Jurassic of the Cotswolds or the Carboniferous of the north length of the Lancaster. It should not therefore pose such serious problems in maintaining a full level of water.

Many small bridges cross the canal between Little Tring and Wendover and we were pleased to find that none of them had been lowered. After a gap of less than a mile without water we found the canal in water again though the level was fairly low. The water was clear and sparkling and there was a slight flow.

The main length of the arm flows through the beautiful scenery of the rolling wooded foothills of the Chilterns. Where gaps occur in the trees there are extensive views over the plain to the north-west from vantage points nearly 400ft above sea level.

As we entered Wendover, we turned up Wharf Street and reached the end of the canal. Water poured from a pipe into a short arm which led off a wide winding hole, weedy but still in good order. With a small rise in water level, it would be readily navigable with a natural site for a marina in the basin. All that seemed to us to be needed was the clearance of the dewatered length, possibly some repuddling and some dredging and weed removal from the rest of the canal. We did not see every yard of the length and there may be greater obstacles but nothing we saw was beyond the competence of volunteer working parties which have already done so much to restore and reopen our canals.

These, then, are the cuts which lead to nowhere, so often the most delightful destination for a canal boat. They are not on through routes which must be cruised to reach the Thames, the Nene or the Trent. They are little waterways to be explored for their own sakes—they are off the beaten track. We were grateful that Slough and Aylesbury are still open to navigation but sad that Buckingham and Wendover are now beyond the reach of *Rose of Sharon*. Buckingham seems to be beyond recall but Wendover—who knows—may still welcome us in enlightened days to come.

We had one more cut to cruise from the Grand Union, the little arm from Gayton junction down to the river Nene at Northampton. Our opportunity came in the summer of 1971 when we made our way down to the National Rally of boats. We had saved up sufficient holiday to spend a few days on the river before the rally and were therefore moored at the junction on the previous Friday night. Several other people had the same idea and the junction was quite crowded so we determined to let go early next morning. At eight o'clock we quietly nosed into the cut and reached the top lock before anyone else had stirred. The first ½ mile consists of permanent moorings followed by a clay pit from which the British Waterways Board take their clay to be puddled for channel repairs. At the top lock two young men were waiting and we learnt that they were canal staff ready to assist the boats through the locks.

The first twelve locks are as close together as those on the great flight at Tardebigge. The locks themselves are narrow gauge and have a rise of little more than 6ft and we found the going excellent. As we stepped down into the valley of the Nene we approached closer and closer to the motorway M1 which crossed over a huge concrete bridge. Though we started in bright sunshine, thunder clouds began to build up and the storm burst upon us at the eleventh lock. Our helpers were on ahead sheltering beneath the motorway and we had just time to erect our cockpit cover and dive below. We were already in the lock but the downpour was so sudden and heavy that no one following was likely to be held up.

When the storm cleared, we pushed on beyond the motorway and found the remaining five locks much more widely spaced. On the flight there are some small lift bridges like those on the Caldon and the Welsh canals but all were raised. The canal soon approaches the river and a junction might have been made 2 miles before Northampton. No doubt this was considered though the river channel is shallow and subject to floods. The last $\frac{1}{2}$ mile is typical urban scenery and the last lock close to the junction where the canal sweeps into the wide river crossed almost immediately by the arches of South Bridge.

We found the little cut delightful, the locks easy and the country attractive except for the last $\frac{1}{2}$ mile. It seemed strange that so small and short a canal should be the only link between the waterway network of the whole of England and the extensive navigations of the Fen rivers. Furthermore, both Grand Union and Nene locks are wide, but this little link allows only 7ft beam and less to pass through. We were thankful for our narrow cruiser but sorry that so few Thames craft could reach the Nene and that few craft on the Nene could go beyond Northampton. Perhaps one day the link will be widened for all to pass.

CHAPTER EIGHT

SORTIE ON THE NENE

MANY have written of the river Nene (Nen or Nyne as Priestley would have it) and most have fallen love with it. For us it is so far from home that we have not yet had a chance to explore it fully. We have decided quite firmly that when I retire we will have a whole summer on the Fen rivers and the linking drainage cuts, but brief sorties are all that are possible in the meantime. Thus the Northampton rally gave us our first chance to explore. We could allow only five or six days and decided that it would be too much of a rush to cover the whole of the non-tidal river to Dog-in-a-Doublet and back. We would therefore go as far as we could but leave time to explore at least some of the villages.

We entered the river via the Northampton arm from Gayton junction and moored to collect our various lock keys. We had written in advance to the Welland and Nene River Authority in Oundle and the keys were waiting for us. Our first lock was a few hundred yards down river and we moored again to prepare it.

We had heard all sorts of stories about the Nene locks and the town lock certainly lives up to expectations. All locks on the river have standard mitred gates at the top and a single guillotine gate at the bottom which must be left fully open when the lock is not actually in use. In this way the upper gates can be used for flood control. The first two locks (Town and Rush

Mills) and one other (Ditchford) have what are called 'radial'-type guillotine gates as against the simple flat vertical sort. They are in the form of a segment or section of a cylinder and they take something like 150 turns of the huge handle to wind down and again to wind up. The handle is locked into a slot and must be returned after use. Turning the handle is not particularly hard, but it is a fit man who carries on to complete the 150 turns without stopping. I was soon out of breath and needed more than one rest.

Below Town lock is the beautiful Beckets Park followed by wide meadows which were to form part of the rally site. We moored to replenish our stores and set off again down river. Having passed Rush Mills, prettily set but marred by a huge pile of derelict cars, we carried on through two more locks with vertical bottom gates which needed only eighty-eight turns of the handle to lower or raise. These, I could manage without a breather. We were now well out of Northampton and chose a place for mooring amongst fields.

I awoke in the middle of the night with a curious sensation—we were tipping up. Had we tilted much more I should have fallen out of bed. The same thing had happened to us once before on the Thames, for rivers have a habit of rising or falling during the night. I quickly put on my thigh boots and a pullover and climbed out to investigate. Sure enough, we were ledging on a mudbank, and our mooring ropes were quite taut. I loosened both, scrambled on to the mud and soon had *Rose of Sharon* floating free again. With her smooth, flat bottom she can be pushed off mud quite easily so long as she is not too firmly aground. For the rest of the night I slept lightly but we remained afloat.

Our first lock in the morning was Clifford Mill and here we met our first snag—it would not fill. Only one of the top gate paddles was working and we could not get a level. Something was clearly ledging beneath the bottom gate and water was pouring out from underneath. I raised it quickly a few turns and wound it down again. I had to do the same thing a second time and this appeared to wash away the obstruction. Some lads camping in the next field helped to push the gates open and we came through without further trouble.

The day was fine and sunny and we had a good run down some 17 miles and thirteen locks past Wellingborough and Irthlingborough with its charming ancient bridge. For most of the way the scenery had been mixed with a number of attractive villages but also wide pits from which gravel and ferric sand had been extracted. We found a perfect evening mooring against a large field with deep water right up to the bank. After supper the farmer and his wife and daughter walked over and came aboard and we had a long and pleasant chat. We found everyone in the country and the villages most friendly and interested in their lovely river, but they all told tales of winter floods and raging torrents very different from the placid stream on which we were cruising. From this point onwards we found the river scenery at its very best.

The next day the weather started to break. For most of the day we had a strong, gusty following wind which made it most difficult to cruise into a lock. Pearl would put me ashore and most Nene locks have very rough approaches often overgrown with nettles and thistles, and I would prepare the lock and open one gate. With no wind, it was easy to slip through this but with strong gusts it was much more difficult. If I had to open the other gate as well, it meant going right round the lock and sometimes crossing on top of the guillotine gate. Closing both gates was just as lengthy a proceeding.

On one occasion when I was steering the lock was filling with a boat inside. This was ideal until I realised that I could not hold back against the wind and current and was too far out in the stream to moor. I executed a nice U-turn, calling out that I wanted the lock and was merely manoeuvring. Unfortunately I could not turn back as the wind was in the wrong quarter, and I came downstream broadside. Luckily we poled ourselves straight just in time and slipped into the lock. On another occasion, Pearl cruised into the lock before the occupant had come out, but she did it so neatly without touching the other boat that its owner clearly thought that this had been intended.

At Denford we bought much-needed paraffin but very little else for the village was very small. We cruised on beyond the ancient bridge at Islip where there is an old and rotting landing stage. We moored to it with some trepidation for it looked as if it might collapse and I hardly dared set foot on it. However, we were able to make our way into the village where we found some good shops which set us up with all we required. A new estate was being built and when completed, the landing stage will be completely isolated and boats will call no more at the village. From Islip northwards, the locks are slightly harder, needing 120 turns, but the scenery becomes more and more charming. The great bend in the river by Wadenhoe and the deep woodlands and fine houses of Lilford present river scenery at its finest. Below the Lilford reach is Upper Barnwell lock and the entrance to the splendid Oundle Marina. We were most amused at this point for a large cruiser shot out of the marina and rammed the bank opposite before it could make its turn.

A little way below the entrance to the marina is Lower Barnwell lock and moored below this was a cruiser whose crew were sitting on the bank having a picnic. The field adjoining was a large one and cattle were grazing in the distance but in the middle, quite isolated, stood a fine big bull. As we came through the lock I called out, 'You do know there's a bull in the field, don't you?' This remark seemed to cause great consternation and as we went on round the bend the picnic party appeared to be in a hurry to reboard their boat!

There is a huge bend in the river at Oundle for it swings a mile eastwards and then curves back and drops through another lock to the town mooring. Thus the distance from the marina to the town mooring is less than a mile by road but almost 3 miles by water. We moored for the night above this lock in very pleasant country. That night and next morning the rain came down in torrents. There seemed little point in going on down river and we were uncertain of the effects of such rainfall on river levels and current. We turned reluctantly and cruised into Oundle Marina where we took in fuel and left the boat to explore the town which we found most interesting. After an excellent lunch at the old Talbot Hotel we returned to *Rose of Sharon* and set off again in fine weather.

We made our way peacefully up the Lilford reach, the rain having apparently made little difference to water levels and moored for tea and I had a swim in deep clear water. While we were moored a large hired boat passed us at full speed, swinging in close to us and nearly washing us off our moorings. Furthermore its hirers passed through both Lilford and Wadenhoe locks without emptying them after use and I had to raise the guillotine gates, lower them and raise them again in each case. Some people at Lilford had told them that this laziness was most improper but had received a short answer. In consequence we were very cross indeed and wondered whether to report them to the River Authority. Actually I had my revenge next day for I found them moored for lunch about 50yd below Denford lock. We cruised into the lock, filled it and cruised out again and I wound up the guillotine gate as quickly as I could manage and shot a lockful of water downstream on to them. I had the satisfaction of seeing them jump for the bank to hold their moorings. All this was rather naughty but I think it had a salutary effect for when they next passed us they came by much more politely.

To return to the previous day, we moored for the night above Wadenhoe lock against a field and opposite the wooded hillside. Next morning I went into Wadenhoe for a few stores and was quite delighted with the village. When we cast off we found the river flowing much faster though the level had not risen noticeably. The first lock we reached was Titchmarsh, one of the deeper locks on the river, and here we saw a rather horrifying sight—there was a good foot of water cascading over the top gates and both paddles were fully open! The lock-keeper told us that a flood was on its way down from Northampton and he was preparing to pass the water through. The water flowed right through the lock and we had to stem the current to get in. No sooner was the guillotine gate down than the lock was full. I was not very happy crossing the top gates with such a raging torrent beneath me, but it all looked much worse than it was. In this lock we were joined by another boat going to the Rally, *Conundrum II*.

Locks above Titchmarsh were rather less fierce though water still poured over their top gates. The powerful currents brought a further hazard, however, in the form of huge rafts of weed above the top gates. I had to rake some of this out and then Pearl would try to get up speed while still in the lock in order to push the weed out into a wider section.

The river is notable for a number of fairly low bridges with a clearance of about 7ft in ordinary conditions. Our forepeak is 5ft 9in above water and we had plenty of room, but with the river 1ft up we had to be careful especially as some foot-bridges narrowed the channel and speeded the current. We were thankful we were coming upstream and not down.

All went well until evening when we approached the notorious Higham Ferrers bridge. This is a brick bridge whose single arch slopes down on each side and is placed on a sharp bend in the river. The channel above is fairly narrow and deep but below it widens out considerably. The actual height in the centre of the arch is given as 8ft in normal conditions, but the conditions now were anything but normal.

As we approached it we noticed three boats moored to the bank and the owner of one called to warn us of the difficulties. He suggested that we should have a good look at it before we made any attempt to go through and told us of two other boats whose cabin tops had been damaged by impact with the brick-work. We moored, looked, and decided to stay where we were.

From beside the bridge we could see that the stream flowed very fast beneath the arch obliquely and fanned out into whirl-pools below. The surface was rippled into small waves and the headroom seemed greatly restricted. We decided to stay the night and await developments and two more boats joined us later.

Next morning was fine and, though the current was still very fast, the level had dropped about an inch on the brickwork on the underside of the bridge. Our friend, who had warned us the previous evening, phoned the River Board and learnt that the level was likely to drop about 3in in the next four hours. We did some shopping in the pretty village and continued to wait. About mid-morning, narrowboat *Grange* came up in the capable hands of its owner Ken Dunham and just succeeded in running beneath the bridge. By now we all knew what the current was doing and *Conundrum* decided to have a go. She, too, came through unscathed. The owner of *Arak Ahoy* was worried that his 6hp outboard would not have sufficient power to push his craft against the current, so we floated a long line down to him and pulled him through. His boat was very light and bobbed about like a cork and he had people on deck with thick gloves and sweaters to fend him off. They were covered with brick dust when they emerged.

At this point a narrowboat with high superstructure was seen coming downstream fast. Despite our shouts he came on and we realised that his reverse was having very little braking effect. At the last moment he rammed the bank and many willing hands took ropes. He was only 10yd short of the bridge when we had him firmly moored and we suggested that he should have come on and carried away the whole bridge to release us all!

By lunchtime we could see two more bricks uncovered by the water and one by one the moored boats set off. We all knew the drill by now and were prepared to cast off ourselves. We went right over to the other side, nosed into the swiftest part of the current and opened up the throttle. *Rose of Sharon* passed slowly but smoothly under the bridge with about 8in clearance.

From this point onwards, although the levels were still high and the current strong, we had no more problems. We moored for the night above Doddington in deep water alongside a meadow and cruised on up to Northampton next day. Our last nine locks were shared with Ron Stainton in a very nice Oundle hireboat, *Merry Enterprise,* and this reduced the work considerably. At Rush Mills we were checked in to the National Boat Rally and given full instructions to cruise on to a good mooring on Midsummer Meadow.

We had found the Nene delightful and had seen it in two moods, the one showing clear, limpid sunlit waters and the other a brown, swirling torrent. We realised that despite all the careful precautions of the River Board, it could have been much worse and that at its worst, even the guillotine gates would not have given sufficient clearance for our boat. Our farmer friend had told us of floods over his meadows in winter and spring with the river $\frac{1}{2}$ mile wide. This kind of hazard can occur on many rivers and we had even seen it from the safety of dry land on the Soar and the Stratford Avon.

We found it a river deserving of time. The locks are numerous and heavy but taken gently they are seldom hard. The villages are beautiful and worthy of thorough exploration and we were told that many of the backwaters used to be attractive but were now too weedy to enter. A week was far too short even to see half the river properly. We should have liked a month on the Nene and the rest of the summer to explore the Middle Level and the Ouse. This we shall keep as a treat still in store for us and shall look forward some day to cruising lazily through the whole length of these delightful rivers.

CHAPTER NINE

THE RIVERS LEE AND STORT

IN 1969 we were able to plan for a longer holiday than usual and the question arose as to where we should go. 'Let's make it London via the Grand Union.' 'Let's go over the top at Foxton.' 'Let's do a round trip, entering the Thames at Brentford and returning through Oxford, taking in the Wey en route.' Not the lower Thames in August, we decided, with its regattas and its queues for the locks. That should be reserved for a holiday earlier in the year.

Then Pearl said, 'Let's try the Lee and the Stort.' At first I thought it impossible in the time, but I pulled out all the British Waterways booklets and worked out a schedule. We could manage it and have five days on the rivers if all went according to plan. Actually we moved more quickly than I had planned and we had almost a week.

'Don't let us tell anyone where we are going until we get there,' we said, for all manner of hold-ups might have prevented us from reaching our destination. 'We'll just say we hope to reach London.' We kept to our plan and enjoyed sending out our postcards from Bishops Stortford and Hertford.

When our holiday started, we came south from our home mooring in Cheshire via the Trent & Mersey canal and the Leicester section of the Grand Union. We crossed the Chiltern Hills at Cow Roast and dropped down the interminable flight of locks to Bulls Bridge, Southall. On a brilliant hot day we

Grand Union Canal: (*above*) Watford locks on the Leicester section; (*below*) Bulbourne junction of the Tring summit of the Grand Union Canal with Wendover arm on the right

Page 118   (above) Guillotine lock in Northampton on the River Nene;
(below) automatic locks at Tottenham on the river Lee

turned left through the beautifully proportioned bridge opposite the British Waterways workshops and started our long journey through London.

The canal from Bulls Bridge describes an arc, running first north-eastwards then eastwards and finally south-eastwards into the heart of the metropolis. It skirts the factories of Southall and Greenford and continues past Lyons, Glaxo and other well-known names, some with most appetising smells. It then separates acres of houses from the attractive rural countryside bordering Horsendon Hill, on whose eastern slopes is the Sudbury golf course. We moored for tea under the trees, 50yd from the fairway.

We had wondered where to stay for the night and this seemed an ideal spot, pleasant and peaceful and only a couple of hours' cruise from Paddington. We determined to have an early night and make ready for a long and arduous day which was bound to follow. As night fell, the sky over London was illuminated with the reflection of myriads of lights in the city.

Next morning we were up at six and cast off at eight for our run into London which included everything that might be expected on an urban canal. There were factories and houses and an aqueduct over the North Circular Road; there was a huge gasworks and extensive railway sidings; Wormwood Scrubbs lay off to the south and the canal skirted a large cemetery; there were terraces of old houses that have seen more prosperous days and giant flats rising to take their place, with demolition and building going on side by side; and through it all, the well-maintained canal runs wide and deep.

The one thing that disturbed us considerably as we cruised along was the scene of destructiveness on the canal side. We saw large, new factories, not even completed, with every pane of glass broken in the windows, the gravelled towpaths providing the ready ammunition. There was a motor-cycle police patrol but he had too large a beat to protect properly. Flinty gravel is the cheapest and most easily available material for paths in the London area but it seemed to us that tarmac or concrete would

H

be more suitable with mischievous or malicious children about. We saw lots of children throwing stones into the water, an action we have all done at some time in our lives, and one child threw stones at us. It is, I suppose, a short step to breaking windows in unoccupied premises. It would pay canal-side property owners in the London area to subscribe to the British Waterways Board for ammunition-free towpaths.

Quite suddenly, we were entering Little Venice. We cruised slowly by the trip boats *Jason* and *Serpens* and the Canaletto Gallery, past the moored houseboats and into the narrows adjoining the office of the British Waterways Board, and there I went to report.

When we had planned our trip, I had written for advice to the pleasure craft officer of the board. I had a long and helpful reply which suggested that I should call at the Paddington office on the way through and this I did. They immediately notified the lock-keeper at City Road locks and also arranged for me to collect some diesel fuel at Camden Town. It was then a few minutes to ten o'clock and we saw the zoo bus taking on passengers. We turned left past the beautiful terrace of Georgian houses and cruised towards the Maida Vale Tunnel.

We were not long through the tunnel when we saw the first of the zoo buses on our tail and he passed us just before Blow-up bridge. This bridge marks the site of a terrific explosion last century when a boat full of gunpowder ignited. The bridge was rebuilt with its original cast-iron supports positioned slightly differently so that the hollows worn by the ropes are now on the inside, away from the canal, where no horse could have passed.

At this point, we entered London Zoo and cruised past antelopes and the vast aviary designed by Lord Snowdon. We had previously done this trip on *Jason* but it was a particular thrill to do it again in *Rose of Sharon*. Ahead of us was the arm which used to lead to Cumberland Basin with its marina and floating restaurant, the *Barque and Bite*, and we made a sharp turn left towards Camden Town.

We cruised past a garden where a little girl called out to us

'Clive Jenkins lives here,' and another with boats moored, belonging to Viscount St Davids. A little farther on we passed the
barge headquarters of the wonderful Regent's Boat Club run by
Viscount St Davids who invited us aboard. As we had a long
way to go, we arranged to visit them on our way back.

The Regent's Canal Society has a scheme for developing the
whole waterway through this part of London which urgently
needs every amenity that can be provided. Bridges are to have
their parapets lowered, towpaths are to be cleared, basins are to
be turned into marinas and the whole canal-side redeveloped
and opened out. Actually, beneath the only bridge we passed
which had a low parapet, we were spat on by small children. I
have already noted the stone-throwing and both these mischiefs
need to be considered by the planners. I wondered how best they
could be solved until I visited Viscount St Davids' club. Then
it was clear that he had the answer.

We had moored at the top of the Hampstead Road locks on
our return journey when one of the many small children expertly
handling canoes and dinghies asked us the name of our boat.
He obviously reported our presence for we received a great
welcome when we cruised up to the barge headquarters. Willing
hands took our mooring lines, helped us off on to a floating
pier and handed us safely aboard the barge. There we met
Viscount St Davids, his captain and a senior skipper and others
and were shown round. The club members, aged seven to fifteen,
are a happy, vigorous crowd from all sorts of homes, enthusiastic
and wonderfully disciplined. They had built their lockers with
salvaged timber and their great boiler, very necessary for drying
clothes, was fired completely by flotsam. One small incident
particularly impressed us. The children had saved a bedraggled
pigeon from drowning and were tenderly watching its recovery.
'It's moving,' they said, 'it will be all right.' Here were children
from the same areas as the spitting, stone-throwing brats we
had already noticed. But the enthusiasm of one man, whose
imagination had caught theirs, had transformed them into
responsible citizens and, at the same time, had replaced their

boredom with purpose. We reboarded *Rose of Sharon* delighted
—and humbled.

Pearl shopped in Camden Town at the top of the locks. This
was also on our return journey and she set off blithely enough
but did not return for a full hour. Lindy and I were beginning
to wonder whether we ought to send out a search party, when
she reappeared. She had started along one road, inquired for a
pillar-box, had been directed to a cartwheel road junction and
had tried to return along the wrong spoke! She had gone some
distance before she realised her mistake and was then completely
lost. Luckily the trip boat *Jenny Wren* starts from below the first
lock and a policeman was able to direct her back to this spot.
How much less confusing the 'cut' is than the road!

Hampstead Road locks start the long slope down to the river
Lee. We took fuel below the first and continued down the next two
assisted by a young man whose boat was moored near by. We then
worked ourselves down the right-hand of the paired St Pancras
locks which was brightened by a little lock garden in an otherwise very squalid area. We then approached the western portal
of the Islington tunnel.

The two ends of the tunnel are amazingly different. Terraces
of houses stand over the western end which is approached on a
slight curve making it impossible to see if the tunnel is clear
until the boat is almost ready to enter. The eastern end, on the
other hand, is pleasantly wooded and is approached by a
straight, tree-lined cutting, giving a clear view through from the
City Road locks. The tunnel itself is high, dry and nearly three-
quarters of a mile long.

All the locks on the Regent's canal are wide and in pairs and
one stood open for us at City Road and we cruised in. The lock-
keeper and a boy closed the gates, raised the paddles and put us
through. I thanked them but the boy said, 'You'll see a lot more
of us yet.' The lock-keeper jumped on a motor-cycle, the boy
on the pillion and off they went along the towpath accompanied
by a large black dog. They went through the bridge-holes side
by side, in one case meeting an unfortunate fisherman on his

cycle. Occasionally, the dog stopped to see off a rival but they were all ready at the next lock with the gates open to receive us. By now it was lunchtime, but it was obviously impossible for us to stop, so Pearl cut sandwiches which we ate as we cruised along. In this manner we passed through Actons and Old Ford locks and turned into 'Ducketts', the Hertford Union canal.

This canal was built in 1824 by Sir George Duckett, Bart, to link this important waterway with the river Lee and to avoid the need for vessels using these two navigations to drop down into the tidal Thames. 'Ducketts', as it was called, is quite straight, 1¼ miles long, level at first before dropping down through three locks to the river Lee. Having turned in, we pressed on for the lock-keeper had already reached the locks and the top gates were open for us. The wide locks occurred singly and were in rather a shaky condition with a good deal of rubbish and floating timber in them. It was for this reason that British Waterways like their lock-keepers to be around when pleasure boatmen go through for a piece of wood jammed in the gates could empty a pound and put the timber barges on the mud.

As the last lock gates were opened, the lock-keeper called to us, 'There's the Lee, mind how you go.' When I thanked him, he asked, 'When are you coming back, guv?' I told him, 'Next Monday at about half-past ten.' 'I'll be around, guv,' and sure enough he was, with the lock already prepared for us. Cruising the narrow canals of the Midlands and the North, we seldom see lock-keepers. In London and on the Lee and the Stort, we saw a lot of them and found them without exception the most friendly and helpful people I have ever come across. It was this, I think, that stands out most in my many pleasant memories of the trip.

It was half-past two when we turned into the Lee and we had already cruised nearly seven hours without a break and the lower Lee is not an ideal place to moor. We were told that it was the busiest commercial river in England and this we could well believe. It is as well that we did not know this before we started —perhaps we should not have come, and we should then have missed some wonderful experiences. We pressed on. By now the day was absolutely scorching and there was no shade. Luckily, we had laid in a stock of beer and Pearl kept my tankard filled.

The first sight of the wide, deep, but gently winding Lee is a huge timber yard with cranes unloading timbers from Thames lighters. We realised that before long we must meet some of these lighters on the move but were quite unprepared for the small tug and string of five laden to the gunwales which bore down on us. We ought, so the tugman told us picturesquely, to have crossed over to the left side to have given him the outside of the bend. In fact we found a tiny hollow in the towpath into which we could squeeze and hold on. The lightermen had some difficulty controlling their craft with their huge tillers and we were glad not to receive a sideways nudge of 100 tons. We cast off and cruised on having learned our lesson.

Our first lock was Tottenham, one of an automatic pair controlled by a lock-keeper sitting in a little glass-sided office. Lighted arrows showed us that we could use either and we cruised into the one on the right. We gave our licence number, the name of our boat and the gates closed for us to rise gently and cruise out into the upper level. It was all so simple and gentle and seemed to take little more than a matter of seconds.

Stonebridge locks, also automatic, came next and then Picketts, a single lock only, which was manually operated. It was beyond here, alongside the Brimsdow Power Station that we had our first real hold-up. We came up behind a tug with a string of lighters which were being cast off, one by one, to drift in to their moorings. We did not want to become involved in their manoeuvres and as we had now been on the move for nine exciting hours, we felt we deserved a short tea break.

It had been our aim to cruise out into the country for our night mooring so we were soon on our way again. We passed through Ponders End lock, the last of the automatic locks and on through Enfield lock where a tug and two lighters were moored. There, we stopped for a drink and some good advice from an ex-boatman on a suitable mooring for the night. This was above Waltham Town lock in a little corner on the opposite side to the towpath where we should be out of everyone's way. We made for the spot and were thankful to stop for we had travelled for eleven out of twelve and a half hours, had covered 22 miles and had passed through eighteen locks on an eventful day. The lock-keeper came to visit us at our mooring and his last words were, 'I shall be up here early in the morning, would you like milk and a paper?' When we thanked him, he commented, 'Oh, it is all part of the service!' Needless to say, he was as good as his word and there were two bottles and a *Daily Telegraph* on our counter before we were up.

Next morning we were awakened by the gentle sound of rain dripping from the trees on to the cabin roof. We were up and having breakfast when the tug appeared, entered the lock, came up in it and pushed open the top gates with its huge fender. The lock was immediately emptied and the tug, with its ropes over the gates, pulled in the first lighter. When this had joined it, the ropes were passed back to the second lighter which was also soon rising in the lock. The tug captain came across the lock to us. 'Thinkin' o' moovin'?' he asked. We had hoped to take our time over breakfast, but—yes—we hoped to go up to Ware later in the morning. 'Then I should start now and keep ahead of us,' he said, 'we've two more barges to collect and may take an hour over each lock.' It was nice of him to warn us and we took the hint.

From this point upwards we were in open country for most of the way. We worked the locks without assistance and enjoyed the scenery of a much more rural navigation though it was not until we reached Ware and pushed on towards Hertford that we felt we were on a river. The Lee has been canalised for so long that it is quite unlike any river we had ever seen before.

The wide Lee valley is gravel-filled and several rivers seem to run in it. I recalled my student days when I learnt that the river Thames had once taken a more tortuous path to the sea, with a huge bend northwards via the Colne and Lee valleys. During the Great Ice Age, the ice sheet had blocked this bend and the river had had to cut a new channel through what is now London. Thus the Lee valley has immense spreads of gravel deposited by a much larger river and the present river meanders through a valley far too large for its waters. Furthermore, the waters in the valley are not one river but three or four. There is the river itself which winds in and out of the navigation but mostly lies away to the east. There is the navigation, the deep channel, which is almost a canal throughout most of its course. There is the 'New River' on the west side of the valley, which leaves the Lee below Hertford passing through a measuring house where the water is metered, to provide some of London's water supply. This is held ready in a succession of large reservoirs also in the valley.

The history of the river as a navigation is a particularly long one. The first Act of Parliament for its improvement was as early as 1424 in the reign of Henry VI and a second Act was also passed in that reign six years later. These were for preservation and scouring the river Lee up as far as Ware. J. Priestley's comments on these Acts in his monumental work of 1831 are rather delightful—'This, according to the custom of the time, is written in the Old Norman French, and therefore need not be related here!' Fortunately W. T. Jackman was able to read Old Norman French and in his *The Development of Transportation in Modern England*, he notes that there were so many 'shelfs' in the river between Ware and the Thames that boats could not pass along its course. The Chancellor was permitted to appoint Commissioners to remove these shelfs with authority to borrow money if necessary and for three years to take a fourpenny toll on each freighted boat. Jackman notes that this is the first instance known to him of tolls charged on rivers to reimburse public work. It further shows the importance of the navigation in those days at least as far as Ware. The sum of fourpence was approximately a day's wage for a labourer at that time and a comparable toll today might be nearer three pounds. I would like to know how dredging and scouring were carried out at that early period.

Jackman again is our authority for the Lee during the reign

of Queen Elizabeth. An Act of 1571 permitted a new cut to be made from the river into north London so that 'barges, tilt boats and wherries could carry goods and passengers between London and Ware'. This cut was to be completed in ten years at the expense of the counties of Middlesex, Essex and Hertford, and the work was carried out. The cut was an undoubted success but it raised the anger of those who had previously carried corn to the city by road and who now did all in their power to sabotage it. First they petitioned against it then, when their petitions failed, they burned the locks, intimidated the lock-keepers and breached the channel. Despite this action repairs were made and carriage by water continued throughout the reign though at times with difficulty.

It is interesting to note the local opposition to navigations at that time even though the Lee had been made navigable over a century earlier. I was reminded of a working party on the Upper Avon recently on a part of the navigation which had fallen into disuse a hundred years earlier. As we cleared the trees and nettles on a piece of unused land, a farmer showed considerable hostility to us and to the whole project.

T. S. Willan, in *River Navigation in England, 1600–1750*, tells us that the lock at Waltham, built about 1574, was a pound lock. Most of the locks on early navigations were flash locks—weirs with gates in them. Passage of these was both difficult and wasteful of water but a few were still to be found on the Thames and the Stratford Avon until a very few years ago. The pound lock in which water is impounded between two sets of gates in a chamber little larger than a vessel, was first used in England on the Exeter canal in 1564. That on the river Lee at Waltham was probably only the second to be built in the country. We wondered whether Waltham Town lock, where we spent the night, was on or near the site of this very ancient Elizabethan example.

In 1608 the New River, already mentioned, was constructed and navigation continued on the Lee itself throughout the seventeenth century. Mr Mottram, lock-keeper at Picketts lock, told us later that Charles II granted the freedom of the Thames to boatmen of the Lee who had continued to bring corn to London during the terrible period of the plague.

The year 1739 saw another Act of Parliament for improving the navigation and yet another in 1767 which included a whole series of new cuts, giving rise to the navigation and locks much as they are today. Twelve years later a further Act made certain additional financial provisions.

Thus the river has a history of navigation of well over five hundred years and is one of the few in England to have been modernised in recent times. Its early pound lock must have seemed as advanced in Elizabethan days as its automatic locks appear today.

As we proceeded upstream, we passed the junction with the river Stort on our right and continued up to the Rye House Inn. It was to this place that the famous great bed of Ware was moved in 1869. The huge four-poster, now in the Victoria and Albert Museum, is said to have slept twelve people at a time. It was also here—but not in the bed—that the infamous Rye House plot was hatched. This was an attempt to murder Charles II and his brother James on their way back from the races at Newmarket. It was in 1683, when the Tories had ousted the Whigs from Parliament and were persecuting Protestant Dissenters, that a group of old Roundhead soldiers planned to waylay and murder the royal brothers. Like so many plots in this period, it went wrong, in this case owing to a fire in the house in Newmarket which forced Charles to leave several days earlier than expected. Though Rye House is not without attraction today, we were disappointed in the surroundings which included a modern factory directly opposite and a stock car racing circuit close by.

The river from here to the town of Ware is much prettier and more rural and we thought Ware itself a charming town. Above the attractive town bridge the river divides and the lock cut to the left is fairly narrow. The lock itself and its surroundings are beautifully kept and we were told that it belongs to the River

Board and not to British Waterways. Though it was only five o'clock when we reached it, the paddles were padlocked and everything appeared to be shut up for the night. There was a beautiful modern house near by where I learnt the lock-keeper was to be found so I went over and knocked on the door. I asked politely to be allowed to work through the lock and was immediately handed the key. While we were in the lock some people commented that we were highly favoured and that the lock-keeper (no doubt rightly) seldom trusted the key to anyone. We were careful to leave the lock exactly as we had found it and returned the key to the lockhouse.

Above the lock, the river curved past well-laid-out gardens and a recreation ground and, looking back, an arm was seen to extend to fine warehouses and well-proportioned buildings, though no barges appeared to be able to reach them now. We moored for the night against the towpath about ½ mile farther up and the navigation seemed here to be much less artificial.

The next morning, still in brilliant weather, we set off up the remaining 2 miles into Hertford. There was one more lock to pass through, with the usual helpful lock-keeper, and we asked him where to moor in the town. 'Oh, your best spot is by the old barge,' he said, so we cruised on towards the town. We passed a good mooring by a public house, crept under a small bridge, turned a corner and squeezed under the town bridge, to the surprise of a number of people who watched us, and came to a full stop. The navigation disappeared completely into a number of small streams which entered the little basin into which we had cruised like the fingers of a hand. We tried to turn, ran aground and pulled out backwards. We managed to wind successfully between the two bridges and cruised back to the mooring by the pub—which we now noticed was called 'The Old Barge'! However, our little journey was not wasted for we like to go as far as possible and reach the head of every navigation.

Hertford is a pleasant town and we spent some time shopping and collecting the mail which we hoped to be lying *poste restante* for us. We had lunch aboard and then cruised back to

Ware passing the measuring house of the 'New River' on the way. This is the prettiest stretch on the whole river and we took our time on it. Back at Ware, where we were planning to meet some cousins, an odd coincidence occurred. On our way from the north to London, we had followed the railway for many miles and had sought out as quiet a spot as we could find for our night mooring. The line carries a great deal of traffic, both by day and by night and trains were roaring through as we went to bed. Pearl woke in the middle of the night to find things strangely quiet. Next morning, we learnt on the early news that there had been a serious derailment in the night at Kings Langley and that all the lines were blocked. Later in the morning we cruised past the accident and saw huge cranes moving upended wagons to clear a single line.

Before we reached the scene of the accident, however, we passed a train which was waiting its turn to get through. We had to lock down in full view of the passengers and this gave them something to watch. In Ware a week later, a lady and a little girl stopped Pearl and said, 'We saw your boat a week ago, we watched from the train while you went through a lock!'

We spent the night moored to the towpath below Ware, a few hundred yards above Hardmead lock. As usual we were up early but were in no hurry to move. During breakfast we noticed activity on the lock and a barge appeared to be rising slowly in it. 'Plenty of time,' we said, 'a tug has got to come from somewhere yet.' What we had not noticed was a very small tractor, built for use on the towpath, standing close by. Suddenly the barge started to move towards us and we then saw both tractor and tow-rope in whose path we were moored. We leapt ashore, cast off and pushed into midstream more quickly than ever before.

Having set off so speedily we continued without a stop to the confluence with the river Stort and turned left, up the river towards Bishops Stortford. At the junction a tug and other lighters were moored, but they were pointing up the Lee and we were not sorry to be out of their way. The Stort locks are rather narrower than those on the Lee, too narrow for the Thames lighters or, for that matter, for a pair of narrowboats. We found this when a cruiser which had passed us at the junction was loth to move far enough forward in the first lock, and there was insufficient room beside him. As there was no shortage of time or water we were content to wait.

Historically, the Stort could never compare with the Lee as a navigational highway. Nevertheless the navigation itself has a history of more than 200 years. The first Act of 1759 permitted the construction of locks, weirs, etc, as far up as the new bridge in Bishops Stortford, but the Commissioners had difficulty in raising the necessary money. According to Priestley, three gentlemen then undertook to complete the navigation and a second Act was passed in 1766 for this purpose. The work had to be finished in five years and this target was handsomely beaten, for we learnt that the bicentenary celebrations were to be held in Bishops Stortford the following month (September 1969). The tonnage rates given show that wheat, barley, malt and flour were the main items to be carried.

A few years after the navigation was open, Whitworth made a survey for the City of London Court of Common Council for a route to link the Stort with the Cam at Cambridge. Finding this to be practical, he recommended the construction of a navigation which would link London with the whole of the Fen rivers, including the Great Ouse and the Nene. The Cam itself and these rivers were already ancient and well-established navigations and a great agricultural and food-producing area would have been opened up for London. Such a canal would have been only 28 miles long with a rise of 80ft to the top level and a fall of 141ft to the Cam. Had an Act been obtained then, there is little doubt that such a canal would have been built.

As it was, the scheme lay dormant for thirty years and an Act was only finally obtained in 1812. The proprietors were empowered to raise £570,000 and a further £300,000 if necessary, but were not to begin the construction of the canal until three-quarters of the money had actually been subscribed. The money came in slowly and a second Act obtained in 1814 permitted a part of the canal to be cut. Less than a quarter of the sum seems actually to have been collected and the canal was never built. John Rennie was to have been the engineer in charge and his high quality work was always very expensive. The high cost was also partly due to the fact that the canal was to have three tunnels, one over 2,000yd long. Its 5 mile top pound was to have been 6ft deep and 16yd wide to act as a reservoir and additional reservoirs were to be built as well as steam-engines to pump water from lower levels. No doubt Rennie recalled that the passage across porous chalk could be hazardous and he determined to have a plentiful supply of water. There were to have been thirty-four locks in all and they would have needed to have been the same width as those on the river Stort.

We regretted that this canal was never built—for it would have made such a fine round trip and would have allowed us to return through Cambridge and via the Cam, the Ouse, the Middle Level and the Nene. Instead, we should have to retrace our footsteps (or perhaps our wake!).

Though their paddles were large and heavy we had found the locks on the Lee easy to pass. The large handles on the paddle gear gave plenty of purchase and the machinery was well greased and much used. The paddle gears on Stort locks were the hardest I have ever used. I should have fared better had my windlass had a longer handle for great effort was needed to raise each of them. Luckily I was in good trim with well over 100 locks behind me, but when I handed my windlass to some friends who joined us they made heavy weather of them. Most of the Stort locks are in open country for the river winds through a delightful valley and we shared some with cattle and one with a couple of most inquisitive horses.

For centuries Hertfordshire and Essex were the granary of London and huge timber-clad mills depended on the river for their motive power. Many of the buildings still stand by the backwaters some little distance from the lock cuts. Flour was an ideal load to transport by water and once the millers were reconciled to the navigation and were satisfied that they would not lose their water power, there was mutual benefit. Furthermore, maltings grew up beside the river. We saw a particularly fine example at Sawbridgeworth with its stacks of brushwood to provide the fuel for roasting the malt.

Malt is the essential constituent of ale and beer and is prepared from barley by steeping the barley in water to allow it to shoot. The actual steeping process is complicated and is carried out at a set temperature with plenty of fresh water and air. The malt or sprouted barley is then dried carefully in special kilns and it is these which may be fired with brushwood. Once more the drying temperature is regulated with great care so as to roast gently but not burn the malt. The malt, so produced is now ready to be steeped again in water to produce a fluid which will ferment into ale.

I had had several swims in both rivers but neither was as clean as I could have wished. The deep draught tugs on the Lee churned the waters and brought up the mud, but we had expected the Stort to be more limpid. We were particularly concerned at the dark, and at times evil-smelling waters in the locks below Harlow New Town. A fisherman on the lockside told us there was heavy pollution and, though a lock-keeper ascribed it to a tug and two barges which had passed through earlier that day, we were pretty sure the water was foul at that point.

Harlow itself makes something of its river, planting grass and trees along the banks which will become very beautiful in a few years' time. Higher up river was the charming town of Sawbridgeworth on the Hertfordshire side of Sheering lock and we moored there for shopping both on our journey up and down the river. The lock-keeper found us a secure temporary mooring out of the way of other boats and we were able to go up past the ancient church into the town.

The next part of our trip to Bishops Stortford and back to Sawbridgeworth was somewhat marred by the weather. Like so many pleasure boatmen, we travel far and try to keep to a schedule, though we always wish for more time to moor and explore. We do not mind wet weather when we are cruising, but crossing fields to the villages and walking round towns in torrential rain can be very miserable. This was one of our really wet days and we reached Bishops Stortford in a downpour. The town itself may be beautiful—we do not know. It turns its back on its pretty little river as though rather ashamed either of the river or of itself. Even a large space which appeared to be a bus station or a car park had so high a wall that we could scarcely see beyond. British Waterways were doing what they could with the towpath but the town itself did not appear to be interested.

We cruised on past a winding hole towards a modern road bridge which appeared to be blocked by a large diameter pipe. We never reached it, however, for we ran firmly aground in soft mud at what was the limit of the navigation. We reversed and pulled back to the winding hole to moor for lunch while the rain continued to fall steadily. After lunch we withdrew and were not sorry to leave Bishops Stortford behind us.

Meanwhile, the tug and two barges which we had passed on our way up were moored below South Mill lock. They were all three bulky vessels and took up the entire navigation except for a narrow channel about 3ft wide. To do them justice, they had no reason to believe that any pleasure boatman would be on the move above this point in these conditions. We descended through the lock and *Rose of Sharon* nudged her steel bows into the gap. I tried to heave the craft sideways, standing like Samson between the columns of the temple but I lacked something of Samson's strength. There was nothing for it but to go ashore and untie the tug and drag it back to make more room. We squeezed through and cruised on down again to Sawbridgeworth.

A glance at the British Waterways booklet shows that the Stort is a navigation which offers very little headroom. The lowest bridges are said to be a mere 6ft 3in above the water. As flooding can occur and reduce headroom still further, we wondered if we should have difficulty in passing the lowest ones. Our height at the forepeak is 5ft 9in, and this allowed us a clear-

Page 135   River Weaver: (*above*) Anderton boat lift linking Weaver and Trent & Mersey Canal; (*below*) site of Pickerings lock showing lock wall, toll-house on right, house once used by manager and later by swing-bridge keeper, and lock-keeper's house in background on left

Page 136 (*above*) Marsden entrance to Standedge tunnel with railway tunnel above; (*below*) Rochdale canal rally cruise with boats entering Dale Street basin

ance of 6in only, but the locks were well under control and we
had plenty of room.

Our last day on the Stort was fine and sunny once more and
we made the best of it. We stopped again at Sawbridgeworth,
cruised gently down the river with the current, had lunch and a
swim in a lovely piece of open country below Hansden Mill lock
and reached Roydon at about three o'clock. There we were meeting the same cousins who had joined us at Ware and they would
gladly have bought the boat from us as she was and where she
was! After a cruise and a picnic tea it seemed a shame to return
them to their car. We chose our night mooring near to the junction.

Next day was Sunday and we cruised on down the Lee. At
Enfield lock we had two examples of 'helpfulness' that we always
try to avoid. I had to fill the lock ready for *Rose of Sharon* and
when the water was level I crossed the top gates and dropped
the paddle on the far side. A helpful young lady, seeing me do
this, dropped the other one as well. The bottom gates leaked
slightly and I had to raise both paddles again before I could
open the top gates. When *Rose of Sharon* was in and the gates
were shut I made for the bottom gate paddles. A very small boy
started to wind one, found it too stiff, and let go of the fixed
windlass just as I had reached out to take it from him. It swung
round and gave me a nasty blow on the forearm. Without the
two helpers we should have been through with far less trouble
or delay.

The automatic locks were being manned on certain Sundays
between midday and 4pm, and we had timed our trip to finish
on one of these Sundays. We had originally planned to moor
above the automatic locks and to come through on the Monday
morning, but it is 8 miles and four locks from Ponders End to
Ducketts. We decided to cover some of this distance on the
Sunday afternoon and the question was—where should we
choose for a night mooring? Although we had cruised up this
section only a few days before neither of us could recall much
of the countryside. Had we done so we should certainly have

I

moored in the area above or below Stonebridge lock which is very pretty. As it was we nosed in about ½ mile below Picketts lock on the offside of the canal.

It was a frightful mooring. The bank was high and when I succeeded in finding a place for the gangplank the whole area was deep in thistles and brambles. I carefully trod a little path through them but Lindy hated it and looked most dejected. She utterly refused to 'be a good dog' and scuttled back on board as soon as possible. Though the river was wide we were within range of the towing path opposite and some boys turned up on cycles and started throwing stones. I stopped them by shouting, but we felt very vulnerable—too close for safety and too far for retaliation. We decided to move back to the tail of Picketts lock. Here we were secure and could take immediate action if anything happened. Needless to say, nothing did and we had a very pleasant evening.

It was not until next morning that we discovered our good fortune in choosing this mooring. The lock-keeper had not been around during the evening but he appeared early and soon we were chatting and we persuaded him to come aboard for a few minutes before we let go. His name is Mr Harry Mottram and he is a poet. Some of his verses have been published and some have appeared in the London Underground Railway advertising wool! He is a most sensitive man, a great lover of the river in all its seasons and moods and a perceptive observer of nature. We were particularly delighted when he wrote one of his poems in the back of my log-book. Had we chosen a better mooring on the previous afternoon we should never have enjoyed Mr Mottram's company or listened to his delightful verses.

We had risen early and had intended to be on the move well before eight. As it was, it was nearly half-past before we let go, but we had the easiest of runs to the junction which we reached at half-past ten. Even so there were one or two anxious moments.

We were back in the commercial area and everything was moving. We were through the automatic Stonebridge and Tottenham locks in quick succession and were then in the thick of things. We passed a tug collecting its barges and approached a bridge partly blocked by two unladen lighters. Before we reached them they blew across and blocked the way. We slowed down while the lightermen shafted them over and then saw that a tug with a string of barges was coming upstream fast. A lighterman called to us, 'You'll 'ave to 'urry if you don't want your paint scratched!' We 'urried! I put the accelerator right up and, in the deep water, *Rose of Sharon* shot downstream through the gap and out of harm's way! Before we reached the junction we met yet another tug with its string of loaded lighters but there was plenty of room. We were not sorry to turn back into the quiet of Ducketts and to see our friendly lock-keeper sitting on the beam of an open lock, ready and waiting for us.

Once more our trip through the locks of the East End of London was made easy and quick for us with every lock ready as we reached it. 'Hand me your rope, guv,' was the cry, and we were brought up smoothly and safely. The lock-keeper still had his motor-cycle and his black dog, but his assistant was missing. This made very little difference to our progression for the lock-keeper leapt across the lock from a closed gate to the one still open. He finally left us at the top of the Hampstead Road lock in Camden Town where we moored for lunch. It was on this occasion that Pearl shopped and was lost and we then visited Lord St Davids' club. Our only other adventure was near the Zoo. We came slowly through the bridge by Cumberland Basin and started to turn right. We were immediately followed by *Jenny Wren* and I had already beckoned it through when my name was shouted from the basin. We could not stop or we should have been run down and we could not moor for boats were all along the bank. We finally pulled in by the Zoo, beneath the footbridge where we could talk to our friend, who had run along the path, through the railings! He was visiting London and had moored his boat comfortably and safely in the basin. This is surely the cheapest and, in many ways, the most comfortable way of seeing the great city.

At this spot the rain started and we had the deep waters of the wide canal all to ourselves. We moored again by the golf course at Sudbury and turned right at Bulls Bridge into the main line of the Grand Union next morning.

This had been a wonderful trip but not one we should want to make very often. London was exciting and the Regent's canal very different from anything we had cruised before. The Lee was even more exciting and very attractive in its upper reaches. The little Stort, too, had its attractions though we wished it had not b een acul-de-sac but had had itsheadwaters linked with the Cam. This would have made perhaps the finest round trip in all the country. But, as I have noted earlier, our lasting impression will be of the kindness and helpfulness of the lock-keepers and water folk we met en route.

CHAPTER TEN

THE RIVER WEAVER

WE had known the Weaver for many years before we had ever cruised upon it. Long before the war my interest in geology had led me to examine the records of well-borings in West Lancashire and in Cheshire. These showed that an ancient land surface was buried deeply with layers of sand, gravel and mud deposited during the Great Ice Age. At that time, the sea-level had been several hundred feet lower than at present and rivers had flowed in deep valleys to a coast lying far to the west. The rock surface over which the Weaver had flowed was nearly 300ft below its present valley floor and the channel had followed a deep, rocky gorge. When the first Mersey road tunnel was being built from Liverpool to Birkenhead some little distance downstream from the older railway tunnel, it was found that the rock channel of the Mersey was less than 100ft below sea-level and was deepening southwards. Thus the lower Mersey had flowed the other way at this period and had been a tributary of the Weaver. All this was long ago when ancient men were hunters and had little bearing on the Weaver as we know it today.

In our many cruises we kept seeing the Weaver. The Middle-wich arm of the 'Shroppie' crosses it on a high embankment with a small tunnel to carry the river through. Looking north from the top of the embankment, we had seen the Weaver Flashes and the white sails on the lower lake. When moored in Nantwich Basin, we had walked up into the town and crossed the river now little more than a large stream. In cruising southwards, we had crossed it again near Audlem, once more on a pierced embankment and there it was even less impressive. We had also seen it from the Welsh canal at Wrenbury, a younger stream still. Traced on the map it follows a semicircular course before it joins the Mersey to flow into the sea.

We had also seen the great Weaver navigation at several points from the Trent and Mersey canal. We had moored at Anderton by the boat lift and had looked across at the ICI works at Winnington all lighted up at night. We had seen it curving away round the hill from Saltersford locks before we entered the tunnel through that very hill. We had followed it past Acton bridge and moored on the hillside above Dutton locks and watched the coasters come up and hoot for the bridge to be swung open. Yet in nine years' cruising we had never made time to explore the river itself for we had always been too busy going somewhere else. As so often happens, however, in 1970 we were on the river three times, first on a three-day cruise in *Rose of Sharon* and twice later in trip boats. Having cruised it, we have now fallen in love with the river and shall return many times in future.

The early history of the Weaver navigation has been written up fully by Professor T. S. Willan in a fascinating book, *The Navigation of the River Weaver*, well worth keeping handy on a cruise. He describes the original eleven locks and gives little plans showing their positions. Many have been removed completely, but it is still possible to see where they were. His book ends at the year 1800 and we hope very much to see a further volume or volumes to bring us to the present day. Hadfield and Biddle give a shorter account in *The Canals of the North West* and this covers the more recent history.

Though there had been several schemes in the seventeenth century to make the river navigable, an Act of Parliament was not obtained until 1721. The main purpose of the navigation and its greatest problems concerned salt. Great beds of salt interleaved with red and green shales and marls underlie much of Cheshire. This can be mined or dissolved out as brine and the salt workers for centuries have done both. Rock-salt is given to cattle and is crushed and spread on icy roads, but it is best purified by solution and evaporation. Brine also needs to be evaporated and in both cases coal is required as a fuel. Coal has always been plentiful in Lancashire and in Staffordshire but carriage on carts or by pack-horse in the early days was slow and costly. Such carriage, particularly of impure rock-salt, had taken place for centuries and place names allow us to trace the old salters' roads. Saltersford, north of Northwich, marked the pack-horse crossing of the Weaver and Salford, near Manchester, has the same derivation. Thus the salters of Northwich and Winsford and even of Nantwich and Middlewich welcomed a navigation which could bring coal from Lancashire nearer to them and even right to their hearths.

On the other hand the mining of salt and more particularly the pumping of brine is apt to cause subsidence. Many of the beautiful meres of Cheshire have been formed where natural brine springs have carried away the bed of salt beneath. Such subsidence is particularly dangerous near a navigation. In the middle of the eighteenth century an entire lock near Northwich subsided into salt workings and within the last twenty years a length of Trent & Mersey canal was rebuilt a few months only before the old course subsided in a similar way.

The Act of 1721 permitted the Weaver to be made navigable from Pickerings, where it was tidal, through Northwich to Winsford town bridge. A further Act, three months later, extended the navigation via the river Dane to Middlewich, but the work was never even started. In 1734, yet another Act allowed the Weaver to be made navigable to Nantwich but this work was never put in hand. Thus, only two of the four salt towns benefitted directly from the navigation and Middlewich and Nantwich had to wait another fifty years for canal links.

Willan describes fully the problems of the commissioners and undertakers in the construction, but the work was finally put in hand and a navigation of sorts was open to Winsford in 1732. Eleven pound locks were built at Pickerings, Dutton Bottoms, Acton Bridge, Saltersford, Winnington, Northwich, Hunts, Hartford, Vale Royal, New Bridge, and Butty Meadows near Winsford. A boat carrying 38 tons of salt could travel from Winsford to Northwich and one carrying 45 tons could navigate the rest of the river to Frodsham. The locks were timber-sided and were already in poor condition by the 1750s. In 1757, the Sankey canal on the north side of the Mersey was completed with locks in its lower reaches 18ft 6in wide. Shippers pressed for the locks of the Weaver to be rebuilt to this width so that the largest flats of the time could carry coal from the St Helens coal-field through to the salt towns. In 1760, the commissioners gained a new Act and set about rebuilding the locks in stone and dredging the channel.

The opening of the Trent & Mersey canal took some of the trade from the Weaver, but this loss was soon offset by the expanding salt trade from Northwich. Furthermore, chutes were built at Anderton so that salt from Middlewich could be off-loaded at this point and reloaded on to Weaver Flats. As shipping was held up on neap tides, a further lock was built above Frodsham and, in 1807, the Weston canal was built to link the navigation with the deep water channel at Weston Point.

Throughout the nineteenth century many more improvements were carried out. A second, larger lock was constructed along-side each lock and, in 1875, the Anderton boat lift was built to make a physical link with the Trent & Mersey canal at a point where the two navigations most nearly approach one another. Then, in the last two decades of the century, the Manchester Ship canal was cut along the south side of the estuary, practically severing the Weaver from all tidal effects. Also, at this time, the eleven locks were reduced, first to nine and finally to four, Dutton, Saltersford, Hunts and Vale Royal. The remaining seven were removed and the whole channel deepened.

We knew something of this history before we started our cruise, and were ready to look out for traces of the earlier locks.

On a bright and sunny Friday morning, we moored *Rose of Sharon* close to the great boat lift at Anderton.

The lift itself has been described by many authors already. It consists of an arm running west from the canal, carried over the valley in an iron trough and two caissons which are entered under guillotine gates. The caissons are iron tanks, each large enough to hold two narrowboats or one barge. Once the boats are in the caissons, these are sealed by gates to prevent loss of water. The end of the canal is similarly sealed. When the lift was first built, it was worked by hydraulic power and the two caissons, each on a hydraulic ram counterbalanced each other. The rams went into cylinders beneath the ground which were connected. The whole system was cunningly designed so that the trough from the canal was a few inches deeper than the arm at the bottom from the Weaver. Thus, when the guillotine gates of the caissons and channels were raised and each was connected to its particular waterway, the upper caisson was deeper by a few inches and was therefore heavier. Loaded boats made no difference as they displaced their own weight of water and it required little more effort than taking off the brake for the two caissons to change places. There is an interesting photograph in the liftman's engine-house showing the lift when it was still hydraulic and the British Waterways Board have plans of it in their Northwich office.

The enormous pressures in the cylinders and against their seals gave trouble and leaks occurred. Thus it was redesigned, each caisson was counterblanced with weights and the entire structure strengthened with sloping metal columns on either side and at the front to carry the extra weight. A small electric motor was put in to drive the caissons.

Having shown us the photograph of the hydraulic lift and something of the machinery, the lift operator told us that all was ready and I was to bring in *Rose of Sharon*. He took my fare and issued me with a large and beautiful ticket with a picture of the lift on the front and full details of its size on the back. We cruised in, made fast and switched off our engine. The guillotine

gates closed behind us and sealed both the caisson and the
Trent & Mersey canal. Our descent started. The gentle movement, the turning of the gear wheels above and the passing of
the great weights half-way down impressed us with the beauty
and ingenuity of the boat lift. Quite soon we reached the bottom
and the whole structure towered above us.

The liftman reappeared, having come down a long flight of
steps and he set to work to raise the gates to let us out. Before
we cast off, he warned us that the gates dripped a bit and then
asked us if we had cruised the Weaver before. When he found
that it was all new to us, he proved a mine of information and
gave us several useful hints about mooring, shopping, etc. We
cruised out and turned upstream towards Northwich.

It was Friday morning, market day in Northwich, and perhaps one of the busiest days for the town. The liftman had told
us to moor against the wall by the swing bridge and this we did.
At the top of the wall was a short cul-de-sac in which everyone
was trying to park cars. Traffic wardens patrolled and a police
car hovered amongst a terrific bustle of people, but we had a
couple of hundred yards of mooring wall all to ourselves. Later,
we shared it with a tug which caused the bridge to be swung
but there was still plenty of room for us both, the policeman
commenting that there was much more room for mooring than
for parking. Across the road were the backs of stores and a passage connected to the town centre. It was ideal for shopping for
everything that we needed and Pearl set off with her baskets.

I had asked the liftman if we should need the bridges to be
swung for us. He took one look at *Rose of Sharon* and replied,
'You could stand on your cabin roof and still get under without
touching.' He was quite right, there was ample headroom. The
bridge in the town is a most interesting piece of engineering for
much of Northwich is subject to subsidence and a swing bridge
with a solid base could easily be thrown out of level and jammed.
Thus the whole structure is said to float in a tank of water whose
level can be adjusted constantly, both to keep the levels right
and to ease the swivelling. When the bridge is across and ready

to carry traffic, it ledges on the two sides which take the strain of the moving vehicles.

We finished our shopping, cast off, passed beneath the swing bridge, cruised by British Waterways yard and reached Hunts lock, the first of the great locks on our journey. Here, we were beckoned into the smaller of the paired locks and we threw ropes up to the lock-keeper. This proved quite unnecessary for he let the water in slowly and we were able to hold the boat in place with our boathooks. We tried to keep off the walls which were muddy and slimy for these locks are now relatively little used. We cruised out above the lock into a very beautiful reach. We were soon away from Northwich, past Pimblotts boatyard where two Admiralty tenders were being built and into a long straight stretch crossed by Hartford bridge. Half-way along, a backwater entered from the west and a number of huge pieces of masonry lay on the bank. On the opposite side was a single wall which had once formed the front of a pretty house. It did not immediately strike us what we were looking at, but we then realised that this was the site of the old Hartford locks.

The river was wide and deep and *Rose of Sharon*, so slow on the shallow canals, was running smoothly and easily at about 6 miles an hour. Ahead of us, on the right, was a parked vehicle and a group of men gazing intently into the depths. One of them waved a blue and white flag and we slowed right down to see what was happening. It was a group of police frogmen and there was a small rubber dinghy, but we were waved on to pass them slowly on the opposite side of the river. Visions of buried treasure, blown safes, and bodies in sacks came before us but the frogmen were only practising. We cruised under the high span of Hartford bridge which we had crossed so often by car. The river continued its straight course for another $\frac{1}{2}$ mile, under a high railway viaduct, and then bore left for the Vale Royal locks, the highest now on the navigation. We were expected and were waved into the narrower, but still enormous lock.

Vale Royal lock was wet and slimy just as Hunts had been and we realised that this is due to the absence of commercial traffic above Northwich. Above the lock for ½ mile or so, the scenery is just as pretty though the navigation still runs straight in the narrow valley whose slopes are so well wooded.

The top 2 miles of the navigation are much less attractive for the banks are lined with saltworks which have now ceased production and the buildings stand empty and derelict. Two channels with a small central island connected to one side by a swing footbridge and the other by a fixed bridge mark the site of New Bridge lock. Some way above this on the right (west) side is the one remaining salt mine in the country, worked by ICI. We saw huge piles of rock-salt waiting to be carted away to corporation dumps, for use in winter on icy roads. Lorries were waiting but no boats were alongside. Though they could collect it so easily there are few places within their reach to deliver it.

I recalled that I had visited the mine before the war with a party of geologists from the Liverpool Geological Society. On that occasion we put our cars in 'Mine Car Park', which to Jennifer suggested the old-fashioned courtesy of 'Mine Host', donned helmets and descended. Rock-salt in Cheshire forms thick seams or layers which are the results of the drying up of inland seas or salt lakes at a period known to geologists as the Triassic. Around these seas some 160 million years ago lived strange, primitive dinosaurs and other reptiles and amphibians which occasionally left footprints in the rippled sands and muds. Fine examples of these ripples in mud, which was also cracked with the heat of the sun and pitted with raindrops, can be seen today in both Manchester and Liverpool museums. The seams of salt are thick and we were able to walk in high galleries, dry and well lighted, and watch modern cutting machines rip out the beautiful crystalline salt. Standing in these surroundings, it is not difficult to picture Cheshire in those dim and distant Triassic days with the hot sun beating down on reddened rock and sand-dunes while lakes dwindled and salt crystallised from solution.

Not far above the salt mine is Winsford Town bridge, the limit of the navigation. Beyond this is a short stretch of winding river, and then the main Winsford Flash. This is a lake, about 1 mile long and ½ mile wide with woodlands on the east side and fields along the west and south. Midway down the western side are the moorings and headquarters of a yacht club, but no one was moving on the lake as we came in from the north. The lake itself was formed by gradual subsidence as underground streams dissolved the underlying beds of salt. We steered east of some 'wrecks' and cruised down the wooded side. Every now and then our propeller threw up mud and we realised that the water was very shallow. We were anxious to moor for tea and Lindy wanted a run on dry land, but we thought we would first try to nose into the channel which connects this lake to the smaller top flash. Running firmly aground, we backed off and made for the south shore with a strong wind behind us. We ran aground again, this time on a hard, gravelly bottom on which we bumped gently. Longing for tea, we went below for a short and rather uncomfortable rest.

Feeling stronger after tea, we took out our two long shafts and poled as hard as we could. Gradually easing off, we went into reverse and ran well out into deep water before turning. Then, heading north again up the west side, we left the flash to the sailing craft and were glad to regain the navigation. We cruised quickly back to the island above Vale Royal lock where Lindy could run to her heart's content while we examined the sluices and the old channel, now a backwater. We asked the lock-keeper if the backwater could be cruised but he advised against it as it had been used as a repository for old lock gates. He had the usual story of a boat whose bottom had been ripped from end to end but thought we might be all right with our ¼in steel. We decided not to take the risk.

The next day was Saturday and, as the locks would be closed from midday until Monday morning, we had to decide where to moor for the week-end. We went down through Vale Royal and as far as Hunts lock to moor and do some shopping and then returned upstream to a length of bank on the west side.

Sunday was a beautiful sunny day and we walked up to the

pretty village of Weaverham. Back at the boat, we traced the old backwater, silted up at the top but still in water below the site of the old lock. Fishermen lined both banks of the river but there was plenty of room for them all. Scullers practised throughout the morning on the long straight stretch which was wide enough for them to keep clear of the fishing lines. In the afternoon Jennifer and her husband joined us and we took them for a cruise in the evening.

Monday morning was just as fine and bright and we cast off early to see as much of the river as possible before it was time to leave it. At Hunts lock we were told that one of the Admiralty tenders was expected as it was to start its trials that day. We had seen it in passing with a lot of smartly dressed men on board, but there must have been a hold-up, for we saw nothing of it throughout the rest of the day. We moored briefly in Northwich for fresh milk and cruised on northwards towards the boat lift and the ICI works at Winnington.

Winnington marks the site of another of the locks removed during the modernisation. Few traces are left except a length of backwater, but there is one further sign which we did not recognise at first. During the last century a series of concrete posts were erected 200yd on either side of each lock, each having the figures 200 cut out across the top. We were told that in bygone days, a boat reaching one of these had priority over other boats farther from the lock on either side of it. In Winnington and at several other ancient locks, the posts are still in place though the locks themselves have gone.

The great locks at Saltersford are very well used and in beautiful condition, constructed of huge pieces of masonry, massive and clean. The Hunts lock-keeper had phoned Saltersford and we were expected. As we cruised into the small lock, still vast in size, the lock-keeper told us about the navigation and gave us a lot of useful advice. The locks themselves are manually operated and the paddle gear on the wide ones are opened by pushing a stout pole round a sort of 'gin'. In the smaller locks the paddles are worked by handles. Lying on the lockside were lengths of worm gear which, we were told, had once been used in the sluice channels to harness the water power to assist in the opening of the sluices. There are also signals on the lockside to direct ships to the prepared lock, but these did not appear to be much in use.

Below Saltersford, we cruised on down the 2 mile reach to the Dutton locks. On our right, a quarter of a mile up the hillside, ran the Trent & Mersey canal, marked by a neatly cut hedge and a number of small bridges. Half-way along we reached Acton bridge on the site of yet another lock now removed. The old course of the Weaver could be seen running to the left with a number of cruisers moored against the island. Willan publishes an old plan which shows a weir a short way down with a small lock beside it, abandoned already in 1774. The lock cut to the right now formed the main channel and we first passed between the buttresses which had once held a swing bridge. Beneath the present swing bridge could be seen the side of Acton lock with the inset masonry for the gate hinge and even the little pigeon-hole steps for the boatmen to climb up the lockside.

The present swing bridge carries the main road over both the main channel and the backwater. This, too, is said to float in a great tank of water which allows ease of swinging and takes up any subsidence. We passed under it easily but were able to watch it swing for larger craft. Looking back up the backwater, we saw the arches of the old Weaver bridge which carried the main road to Acton. We continued down the river towards the Dutton locks.

The Dutton locks are very similar to those at Saltersford, the two locks of different widths lying side by side. Each can be divided so that only part of the lock need be used, with a consequent saving of water. It seemed to me that the wastage of water must be a real problem for the locks are so large and the Weaver so small above the limits of navigation. I was assured that this was rarely so for the Weaver valley has a wide catchment area. In any case, we were given the whole length of the narrower lock and were let down quietly to the lower level.

Below the lock, the backwater comes in from the right side and is crossed by a very beautiful footbridge. It looked a charming spot to choose for a night mooring out of the main channel. A little lower down we cruised under the high Dutton viaduct, which carries main line trains northwards to Liverpool from London, and found ourselves in a most charming stretch of river. The valley is deep and wooded with green meadows on either side of the river, strongly reminiscent of some of the prettier parts of the Thames. In fact, if it had not been for the artificial banks, we could well have imagined ourselves on that river.

We next came to the site of Pickerings lock, once the lowest on the navigation. The locks have been removed, though the side walls are visible on the right bank, with the recesses for the lock gates and the cast-iron pigeon-hole ladder. The lockhouse is a beautiful building, now privately owned, but the backwater which enters on the left side is practically silted up.

Below Pickerings the scenery is just as beautiful and remote for another 3 miles. We saw the cut to Frodsham on the left and ½ mile farther on the river itself swings to the left to tumble over a weir on its way to Frodsham Bridge. The Weaver canal continues straight on to the docks at Weston Point. We turned back before reaching the Weaver canal, which we were to cruise later in a trip boat. We recalled John Seymour's wonderful account in *Voyage into England* of his cruise through the now abandoned Runcorn & Weston canal and up the Runcorn locks, the last boat of any size to have made the journey. If more boats had cruised that route before it was finally closed in 1965, it would still have been possible to follow it today.

Our journey upriver was as pleasant as the run downstream and the lock-keepers as helpful. The current is very slight and it seemed to make little difference to our cruising speed. At Dutton lock, a large coaster appeared but made no wash as we filmed it passing by. Back at the Anderton lift we were expected once more and were raised to the level of the Trent & Mersey canal with the trip boat *Lapwing*. She uses the Weaver and boat lift regularly and takes many people through the beautiful valley.

That night we moored on the wide at Billinge Green against the canal towpath happy to think that we had at last cruised the Weaver and quite determined to explore it again at the earliest chance. It seemed to have so much to show us, from beautiful scenery and great feats of engineering to signs of its long and eventful history. In so few miles it had both industry and rural remoteness and there was room on its waters for ocean-going ships as well as our own little shallow-draught canal boat.

K

CHAPTER ELEVEN

TUNNELS AND LOCKS

THE great canal tunnels vary in appearance from dark holes in the ground like the north end of Harecastle to the wide arches at Netherton. Some are enhanced with whitewashed cottages as at Preston Brook or the south end of Harecastle. There are attractive symmetrical tunnels which you can see through and odd crooked ones which you can not. Some lie a few fathoms only below ground and others, like Dudley, penetrate the high hills. There is nothing in this country, however, which gives a greater impression of burrowing under a mountain than the great Standedge tunnel on the now abandoned Huddersfield Narrow canal. From the western, Lancashire, end the escarpment of Standedge towers above the tiny brick arch of the tunnel which lies alongside Diggle station. On the eastern, Yorkshire, side the tunnel is stone built and, with the two railway tunnels, enters the bottom of a bowl-shaped valley while the Pennine road climbs up to a pass nearly 1,300ft above sea-level.

I was fortunate enough to make a trip through Standedge tunnel in the autumn of 1970 in a maintenance boat—a small ice-breaker incongruously powered with an outboard motor. This is, of course, the easy way, for in early days the tunnel was 'legged' and more recently boats were propelled with shafts, the crew walking the length of the boat with the shaft firmly dug into the side or the roof. In either case, the journey must have seemed interminable. The tunnel is over 3 miles long and once we lost sight of the entrance an hour and a half was to pass before we saw the point of light at the other end.

We met at the British Waterways Depot at Marsden, reached from the Pennine pass by a narrow lane which drops down into the valley. The depot itself is a most interesting late eighteenth-century building and there we were given a briefing by Mr Freeman, the engineer in charge of the north-western canals. He told us how it had been built between 1794 and 1811, how it cost the huge sum of £123,803, how they checked the levels by flooding the workings, how the last loaded boat passed through in 1921 and how three men worked each boat, two legging at a time. He also fitted us up with hard hats which immediately suggested bumped heads.

When all was ready we went aboard the small ice-breaker, lighted the tilley lamps and pushed off. The entrance is quite a small hole, dwarfed by the two railways tunnels, one on either side, both at a slightly higher level. The presence of the canal tunnel had helped the building of the railway tunnels both in providing an easy way of removing the spoil and in draining away the water. It is still essential that the canal tunnel should remain in good order for if it collapsed they too would be in danger of falling in.

The first few hundred yards are arched masonry and this was the original support of the tunnel when such support was needed. Soon, however, we came into bare rock, the massive gritstones of Millstone Grit, so strong that it looked everlasting and we wondered how the tunnellers had ever achieved a passage. We saw the marks of the drills where holes had been bored for explosive and I wondered how many men had been maimed by shots which had blown back.

In the roof were cast-iron plaques every 50yd numbered from the western end, 109 in this great tunnel. We moored for a few minutes to inspect one of the many side passages which climbed up to the southern railway tunnel and then walked on a timber bridge over the canal to the northern, used, railway tunnel, keeping well back in case of express trains which roar through from Liverpool to Leeds. Today they are all diesel, but we could picture the days when they were steam and acrid smoke belched through into the canal and on to the unfortunate boatmen legging their long journey. We returned to the boat and continued our passage.

Where the tunnel is bare rock it is quite high. It is also fairly wide in most places, though in one section there were only inches to spare on either side of our gunwales. We were careful to keep our hands off the sides for our steel boat kept grazing the walls. So long was the tunnel that there were four passing places, great caverns of considerable width and height. We were told that a man used to go from end to end of the tunnel over the top to control the traffic but that these passing points were still necessary. They were called exciting names such as Brunclough, White Horse, Judy and Redbrook.

There are thick beds of shale within and below the gritstone and long lengths of the roof had been strengthened with brick arches when the railways came. Much of this was, in fact, typical railway architecture sometimes reaching below water level and sometimes built into the walls a few feet above the surface. Most of the original building had been done from shafts sunk from the hilltop and headings run out each way until they joined up. We later saw the piles of spoil on the surface. Many of these shafts had been closed but some were still open for ventilation and down them water dripped, one in particular being a veritable waterfall. I was in the bows and turned as I reached shelter to see the misery on the faces of the others as they came through the shower!

At long last the point of light at the far end became visible and we crept slowly towards it. The last 200yd formed an extension built in railways times to carry the station at Diggle and the main line across the canal. We came out through quite a small arch which looks such a miserable hole from the train and climbed into a waiting Land-Rover to take us back over the hill to Marsden. I think that the thoughts of all of us were on the men who had carried out the great task of burrowing through the heart of the Pennines.

By a coincidence I was to have the opportunity to explore the Dudley tunnel a few weeks later. Here the Dudley tunnel trust runs a narrowboat taking trips from end to end and keeping the exits as clear as possible from the rubbish thrown in by vandals. Unfortunately vandalism is high in the area for it is densely populated but the great hill above the tunnel is wooded and beautiful. The society publishes a well-illustrated booklet which gives a detailed history of the actual building which began in 1785 and was completed in 1792. It describes how over one hundred boats a day passed through in the 1850s, either by legging or by shafting, and how this led to the building of the much wider two-way tunnel at Netherton to relieve the traffic.

The hill on which stands Dudley Castle and Zoo is geologically most interesting. Limestones and shales are folded sharply into a dome with steeply dipping sides and to the south-west of this lie the more gently sloping coal-measure sandstones and shales with important workable seams.

We arrived by coach at Parkhead to find the boat awaiting us. We had a quick look at the locks and intervening pounds which had had a tremendous amount of volunteer clearing in the two years since our last visit by boat, then went aboard. Our method of progression was by shafting and a certain amount of amateur legging. The south-western portal was rebuilt in 1884 owing to mining subsidence but our guides from the society told us that there were no signs of further subsidence in recent years.

Most of the length through the coal-measures is bricked and we were told that the bricks were laid in puddled clay up to about 2ft above the water. Above this, loose rock surrounded the brickwork to assist the drainage and slots or weep-holes occurred in the brickwork on both sides every few feet. Through these the strata could be seen.

A dolerite sill, a thick wedge of ancient volcanic rock which had cooled below the original surface, occurs some distance from the entrance and the tunnel opened out to a fairly large cavern as it cut through this rock.

We were past the centre and could see the daylight ahead when we first entered the limestones. Here, once more, the tunnel was unlined and we could see the steeply dipping pale grey strata. One shaft, sealed at the surface, had a most perfect curtain of stalactite.

We passed a wide open shaft and soon came out into Castle Mill Basin where we moored at a small landing stage and went to explore some of the other caverns in the limestone. Castle Mill Basin was once a great junction with a tunnel running over $\frac{1}{2}$ mile under Wren's Nest and another leading south. We scrambled down into one of these, abandoned in the 1830s, and there Dr Ian Langford showed us the sunken remains of an ancient boat. It is hoped to recover this boat in due course and preserve it. I was particularly interested as it seemed to differ considerably in construction from the 'starvationers' which worked the underground mines at Worsley.

Returning to our boat, we had two more short tunnels to pass. Both were bricked and in good condition though the entrances and exits were shallow through fallen rock and great efforts of heaving, shoving and rocking the boat were needed to get us forward. In due course we moored at the Tipton end. Though I was told that the lowest part was only 5ft 9in above water-level I did not get the impression of a very low roof. The whole of the brickwork seemed in excellent condition and the caverns were most exciting.

Another tunnel I should love to explore is Sapperton on the Thames & Severn canal. This canal had great promise, linking the two largest English rivers and had been suggested well over one hundred years before it was actually built, but was dogged by misfortune as leaks developed where it passed over the Cotswold limestone. Furthermore, it was doomed to failure owing to the poor condition of the Upper Thames throughout the nineteenth century. The anonymous author who made the cruise from Manchester to the Thames by way of the river Severn and the Thames & Severn canal in 1869 described weed so thick that it would almost bear a person's weight.

The westerly facing scarp of limestone reaching 1,000ft in elevation and forming the main ridge of the Cotswold Hills, is the watershed between the two river systems. The Severn lies in front of the scarp and the Thames rises on the gently sloping rocks behind it. The line chosen for the canal was from Stroud which was linked to the Severn by the Stroudwater canal, eastwards up the Golden Valley and through the beech-covered hills by Sapperton. The summits at this point are a little over 600ft but this still required a tunnel, believed at the time to be the longest navigable tunnel in the world. Just under $2\frac{1}{4}$ miles, it was exceeded later by Standedge but was designed as a full width barge tunnel.

The first of many good descriptions of navigating the tunnel was by Thomas Love Peacock in *Crotchet Castle* (1831). The party occupied four boats, one for the men, one for the ladies, one for the servants and cooks, and one for the dining-room and band. He wrote of the tunnel 'that the greatest pleasure derivable from visiting a cavern of any sort was that of getting out of it'. Forester, in his novel *Hornblower and the Atropos*, gave a good description of legging through and Temple Thurston in *Flower of Gloster* (1911), actually did the legging himself. The description I like best for the lyrical beauty of the prose occurs in *The Heart of England by Waterway* by William Bliss. It was in the 1890s when he and his friends camped in the woodlands by the western portal and his words conjure up the rare beauty of that delightful spot in summer with the full moon and the nightingale.

We had an autumn weekend in Cirencester and determined to see as much as we could of the canal and its famous tunnel. We found little left of the arm at Cirencester itself and only dry trenches at Cricklade. We then went on to Thames Head bridge on the ancient Fosse Way and saw the remains of the old bridge with a plaque on it giving the dates of the canal (1789) and its abandonment (1927) and the fact that the Gloucestershire County Council had 'realigned' the road. This had made it possible for cars to go so fast that few passengers would notice even the countryside let alone the old bridge.

Next we went to Coates and photographed a fine example of a Round House by what looked to have been a stop lock. Here the lock-keeper had lived and kept watch for boats as they came from the tunnel or from the Thames. Then we saw the beautifully painted sign to the Tunnel House Inn. This is an old canal inn built originally for the men who dug the tunnel and then used by the leggers to restore themselves after their long spells in the darkness. A little lane runs along beside the canal now in a deep cutting. It swings left to the inn over the tunnel entrance itself. We climbed down to examine the eastern portal, still in good condition, with niches on either side said to have been for Sabrina and Father Thames though these were never occupied. The water was clear as crystal but very shallow and quite a current flowed out from the tunnel mouth. I recalled stories of people exploring by canoe and even on foot though the water looked too shallow for the one and too treacherous for the other. In any case the roof has fallen and the tunnel is blocked. After a visit to the Tunnel House and a pint beside its roaring log fire we walked up the old horse path through the woodlands and then set off by car to find the Daneway end.

From the road leading towards Sapperton, we saw man-made mounds crowned with beech trees where the spoil had been taken from the shafts when the tunnel was being built.

A lane runs steeply down to Daneway and to another charming canal tavern built with the tunnel, where navvies slept and leggers waited. Just below it is the last but one lock and the basin where boats waited their turn to make the passage or off-load goods for Sapperton. This is well described in Household's excellent book *The Thames & Severn Canal*. We set off along the overgrown towpath towards the tunnel and in about 500yd reached the western portal which is now rather damaged. Several stones from the top courses of limestone had been pushed over into the channel but it would not be difficult to lift them with a small hoist and re-cement them into place.

Once more I climbed down into the water of the tunnel and listened to a perfect answering echo which seemed to sound from afar. So beautiful was the scenery at both ends of the tunnel even in late November that I could not help feeling the sadness of letting the canal fall into disuse. How great a restoration would it be? Parts have been completely obliterated and built over and long lengths are without water. Leakages both in the limestone slopes to the east and in the tunnel itself have always proved serious problems but modern materials and methods make many things possible. One day quite soon as more and more people discover the beauty of our countryside as seen from the quiet waters, even this size of restoration may be contemplated if what remains is not allowed to vanish still further.

One more series of tunnels must be mentioned here as I was fortunate to explore them while they were still open. These are the underground tunnels of Worsley which link with the Bridgewater canal in Worsley Delph. I had wanted to explore them for some time and a party was finally arranged by Frank Mullineux, the leading authority on this wonderful series of caverns. Four of us met at the offices of the National Coal Board, put on overalls and hard hats, and entered a cage in a mine shaft. This brought us down to the level of the main canal which follows a north–south course below ground from near Bolton to the Delph at Worsley. Not far from the bottom of the shaft was the famous underground inclined plane which linked a canal at a higher level to the main channel and allowed boats to carry coal from the upper canal through to the open air at the Delph. We examined the plane and walked up to see the locks at the top. We then came down again and boarded two 'starvationers', the boats which were used in the underground workings. The main canal was now used for drainage only and the boats carried the inspectors and maintenance men through the whole system. Each was propelled by two men, one at each end, who pushed against the sides and roof with short shafts or lay down and used their toes in the lowest sections.

The main canal was brick-arched for much of its length but ran through solid rock in places. At intervals side passages branched at right angles to the east and the west along the strike of seams of coal, but we kept to the main channel which would eventually have brought us out at Worsley Delph. We actually came to the surface by another shaft.

Some years later the canals were abandoned completely and the water level raised to seal in gas as the pits themselves were no longer worked. A few of the smaller starvationers were brought out into the Delph and I was able to acquire one for Manchester Museum from the Coal Board. The name is said to originate from the fact that the boats were very narrow and all their ribs were showing. Mr Harrison, who had used this particular boat for inspection for many years, caulked it for us and supervised the crane which the Coal Board had kindly lent to put it in the water. The boat is 30ft long, 3ft 8in wide and has vertical sides 2ft high. Pointed at both ends, it is very strongly made, double-bottomed of oak and with oak sides the lower halves of which consist of single planks bent into a curve at both ends. Many years of bumping against the hard rocky sides of the tunnels had marked and dented the timbers, but Mr Harrison told me that one of his gang had known the craft for fifty years and it had not then looked any different. Though we do not know the actual age, it could be anything from 100 to 200 years old.

Our start from Worsley to bring the starvationer to Manchester was unusual—we went off backwards! This was intentional as the press photographers wanted to take us against the background of the black and white Packet House and steps from which passenger boats used to start for Manchester and Liverpool. Having manoeuvred satisfactorily for this purpose, we were about to set off in earnest when our friends from the Coal Board arrived to tell us there had been some slip-up in the arrangements with the police who were to accompany the boat on its low-loader journey along the last mile to the museum. I had to leave the boat for the first part of the trip to sort things out, but everything was soon settled. This was apparently full of drama for we had set the outboard too far forward and the boat would not steer. It would go straight ahead but the canal bends and the journey became hazardous. Then the crossbeam twisted and snapped and Dr Hyde, the reserve steerer, just managed to grab the engine as it plunged into the deep. When I rejoined the craft, the engine had been repositioned and everything was going well. We passed some barges whose steerers were delighted to see so ancient a vessel afloat, crossed the Barton swing aqueduct and carried on to Castlefield. Despite her age and rough life, the starvationer took very little water and our bailer was seldom needed.

Once in Manchester another crane lifted her on to a low-loader and yet another placed her in the museum where she formed the centre of a waterways exhibition. Two other similar boats are preserved, the one at the Monks Hall Museum, Eccles, and the other in Worsley. They are important as the last surviving examples of what was perhaps the very first craft to be designed for a British canal.

I have intentionally left the description of the actual underground canals until last for others saw much more of them than I. By chance quite recently, my attention was drawn to a French bookseller's catalogue which listed a book by two French mining engineers, H. Fournel and I. Dyévre, entitled (in French) *'Memoir of the Underground Canals and Coal Mines of Worsley, near Manchester'*. These two had visited Worsley in February 1842 and had made a thorough inspection of the canals, the workings and the boats and their ninety-six page memoir is in fullest detail. It is so complete that a good model-maker could construct to scale the boats, the inclined plane, the tubs and baskets and the sluices and other mechanical apparatus. They were actually present when the lowest level was being driven and saw the method of working.

The memoir first tells of the coal seams themselves, fifteen of them dipping to the south with the main lines of the underground canals running north and crossing them all. Faults in the rock occur and these bring some of the seams down so that the main canal actually crosses coal twenty-five times. Not all the coal was of good quality, but where it was worth working side canals were driven eastwards and westwards through the coal itself. The coal was then worked up the gentle dip slope from the canal and the canal was used both for the carriage of coal and for drainage.

Not content with the one main canal with all its side branches totalling in all 18 miles, a higher canal was driven practically along the line of the main level and this too had many side branches, in all exceeding 10 miles. Furthermore, when the two Frenchmen were inspecting the works, a third canal at a lower level was being driven and this too had side branches. The total lengths of all canals eventually constructed below ground have been variously given as 43 and 46 miles! The Frenchmen considered that they were the most extensive underground workings in the world. The main level was 8ft high, 9ft wide and the water 3ft 7in deep. It was bricked most of the way but was left unlined where it passed through coarse grits, the bricks sometimes in single and sometimes in double layers. Wider passing places for boats were constructed.

The great inclined plane is described in full detail. It was built on a thick stratum of grit sloping at an angle of one in four.

At the top were two locks 52ft long and 8ft wide, separated by a 3ft wall. The top gates were hinged and had sluices in them and other sluices emptied the locks into a channel at the lower end, the water flowing down into the main level canal. The gates at the lower end were guillotine. As the water ran out under full boats, the boats subsided on to carriages which were 30ft long and had four wheels which ran on rails. A cable round a huge wheel at the top was attached to the carriage in the lock and to another carriage at the bottom of the lift on to which could be floated an empty boat. When the loaded boat was pushed out of the lock on to the incline it pulled the empty boat up, the speed controlled by a brake. Two sets of lines ran side by side down most of the lift but they joined 57yd from the bottom. An empty boat weighed about 4 tons, the carriage 5 tons and the coal up to 12 tons, making some 20 tons in all. Thirty boats could be transferred in this way in an eight-hour shift. The lift was working from October 1797 until about 1826, but was abandoned and derelict in 1842 when the two French engineers saw it.

Apparently the alternative method of raising coal to the surface from the upper level was found to be more convenient. In this tubs were filled with water from the upper canal and these were lowered down pits into the two lower canals by cables passing over drums at the pithead. On linked drums were smaller tubs full of coal which were raised to the surface. The water tubs had special valves which emptied them automatically at their journey's end so that they could be raised to the higher levels and start again.

The third, and lowest, canal level was built to a smaller gauge. The height was only 7ft, the width 6ft 6in and the depth of water about 2ft 9in. The coal from the canals of this level was raised to the middle level to reach daylight by the main canal.

The method of working the coal is very fully described. The miners drove two galleries up the dip of the coal and then ran further galleries to the left and right. Working from the top back towards the canal they took out the coal, supporting the roof with pit props. They did not take coal from within 20ft of the canal so as to maintain the canal walls. The large pieces of coal were then picked out by hand and the rest riddled in 30in diameter sieves with ¾in holes. This gave coal which would not pass the sieve, smaller pieces which dropped through and coal-dust or slack. These qualities were kept separate and were loaded into rectangular baskets by children. The baskets had sledge runners on them, shod with iron, and held 200 to 300lb of coal. One child then dragged the basket and another pushed behind or held it back. In the lowest canal were tub-boats each fitted up with six tubs or barrels approximately 3ft in diameter and

1ft 10in high. Each tub would hold 600 to 700lb of coal. The coal from the baskets was then carefully loaded into the tubs and, when all were full, the boat could go to a pit bottom and the tubs could be lifted by cable to the middle level where the coal was tipped into waiting boats.

The boats are also described in detail. Three sizes occurred, the narrowboats working the side branches of the middle and upper levels, the M boats which were the largest and worked the main line of these levels, and the tub-boats which were constructed on a smaller scale altogether for the lowest canal. The tub-boats were about 30ft long and 4ft wide—ours then was one of the smaller tub-boats. The narrowboats were about 45ft long and 4ft 6in beam and the M boats were 50ft long and about 6ft beam and held about 10 tons of coal. In 1842 there were said to be about 150 M boats and narrowboats in the underground canals and 100 tub-boats. The M boats came out on to the Bridgewater Canal, were attached three abreast, and three groups of three were drawn by horses to the wharfs of Manchester. In this way some 90 tons of coal would be delivered at one time.

The method of working the boats in the canals is also fully described. In the side canals legging was used and this is expressed as being very tiring. In the main canal the boats were collected and brought in groups to Worsley Delph by a most ingenious system. Sluices were lowered to cut the canal into a number of pounds and water gradually built up behind those farthest from the exit. The 'halers' raised the sluices one by one as they reached them, leaving them up and causing a slight current. In this way six men could convey forty boats at a speed of about ½ mile an hour. They timed the travel to bring the boats to Worsley at about 4am! They then legged back empty boats, or boats carrying stores, tools, bricks, etc, dropping the sluices after them.

Though there was a certain amount of gas about, the men worked with lighted candles, the movement of the boats serving to ventilate the pit. Two small jets of gas issuing from the rocks were lighted, one near the exit and the other above the inclined plane. They did not give much light but formed landmarks for the halers.

The whole description of the slow currents silently bearing the boats, the lighted candles, the tiny beacons of flaming gas, the incredible complexity of tunnels and the dripping waters paints a picture of a strange underground world which was surely inhabited by ghosts as well as the halers who worked the boats and the children who filled the baskets.

All these are now closed for the pits they drained are working no more. I was thrilled to have seen them but perhaps it is best now to leave the ghosts in peace.

Leaving tunnels for a while, it is worth giving some thought to locks, their problems and their operation.

When we cruised down the Grand Union canal in 1969 we were interested and delighted to find parts of it so busy. The helpful lock-keeper on the Buckby flight told me that more than three hundred boats a week had been passing through his locks and we met one or more boats coming up through each Braunston lock on our way home. In 1971 we were concerned to find that gate paddles in the top gates of the Buckby flight had been removed and this slowed up locking and caused some congestion.

The pleasure boatmen, particularly if they have just hired for the first time, can be amusing, and sometimes infuriating in their use of locks. They are often in a tearing hurry and quite incapable of looking to see if there is anyone coming the other way. We have waited 30yd or 40yd back from a lock for the boat to come out, only to have the gates shut in front of us. Perhaps the most annoying is the lock wheeler who fills the lock as you approach from below, even though his own boat is not in sight. We had an example of this on the way up to Cow Roast. Gates which had been open suddenly closed in front of us and in spite of loud blowing on our horn, the lock was filled. We moored and Pearl went to see what was happening and found one young man but no boat. He had looked, so he said, but had not seen us. No, he hadn't heard our horn. He was very sorry but did not offer to empty the lock again. I had a word with him and pointed out that he must not do that to working boats, but he seemed frankly incredulous that there might be any working boats. It was with some satisfaction that when we reached Cow Roast we found that a working pair was only ten minutes behind us. We hoped he had learnt his lesson before he met them.

Perhaps the oddest example of lock 'blindness' was on the Hillmorton flight which we were descending. The locks are in pairs and all were empty. We dropped down the first two and were pleased to see a boat coming out of one of the third pair of locks. They left the top gate open and I walked up and leaned on it as *Rose of Sharon* approached gently about 100yd away. Suddenly from below the lock a man appeared and started to wind a bottom gate paddle. 'Here, what are you doing?' I called. 'Oh, I'd better shut the top gate first, hadn't I?' he replied. *Rose of Sharon* was now about 50yd away. I pointed this out and said that we were using the lock and suggested that he might consider the other one which was already empty. He then called down to his friend in the boat below, 'We've got to use the other lock, there's a boat coming.' He then put up fully both paddles in the bottom gates of that lock though it was already completely empty. If I had not been leaning on the open top gate of the full lock we should probably have had the ridiculous situation of a man emptying a full lock to go up while I filled an empty one to drop down!

Several people seem to embark on a holiday with a definite object in view but no idea how to achieve it. We approached the second lock in the Braunston flight to find two boats about to come up in it while we and another boat wished to go down. Finding that one of the boats in the lock was making for Stratford-upon-Avon, we pointed out that he was heading in the wrong direction. Both the man and his wife had climbed out of the boat which was bobbing about near the top gates and the only occupant appeared to be a baby. We refused to allow the paddles to be opened until someone went back on board and made the boat safe. Finally when the lock was full, the boat came out and turned and came down again with us. We explained the route in detail to them but could not help wondering how they would fare in the twenty-one locks of the Hatton flight. It seems strange to find anyone cruising the canals without the appropriate volume of the British Waterways booklets on the canal system. When we use canals not covered by these booklets we make sure that we have the relevant 1in Ordnance Survey maps.

Yet another man was making for the tidal Thames with a suspect engine and a seven-day canal licence. He asked us how far he had to go, how many locks there were and how long the journey might take him. We were in the long flight of locks below Cow Roast and I was relieved to see another boat join him to help him on his way.

While on the subject of locks it has always surprised me that there is so little information on some of the finer points of locking. Every book of instruction tells us which paddles to draw and some of the more complicated locks have instruction boards beside them with little diagrams showing precisely what must be done. Apart from Hadfield and Streat's *Holiday Cruising on Inland Waterways*, there is far less information on the hidden hazards that some locks present.

Dropping down a lock whether narrow or wide should present few difficulties. The boat must not lie too far back and catch the rudder, the skeg or the outboard on the sill. There should be little danger of this for any but narrowboats unless the lock is being shared. Bringing *Rose of Sharon* alone through a deep lock when Pearl had gone shopping, I was once silly enough to let the ropes drop in. The boat seemed a very long way down but I managed to jump on to the cabin top. I could have refilled the lock provided I had done it very slowly and carefully, but this would have wasted both time and water. In some locks I could have flushed her out by opening the paddles but Trent & Mersey ground paddles tend to throw the boat up towards the top gate. A common error is to fail to notice that a gro und

L

paddle above the lock is not completely down and that the lock will not empty. It should be a drill to check top paddles before the lower ones are raised.

A boat that is slightly too wide can jam as the lock empties but the owner will probably know of this danger. I once met a hireboat, however, that could not pass a lock near its base and I have seen a commercial boat jam in one of the Tyrley locks on the 'Shroppie'.

It is in the uphill journeys that catastrophes are most likely to happen. The narrowboat in a narrow lock has little problem for it fits in nicely and has little room to move. Smaller boats have a choice of positions and it is worth considering which is the best place. When I first started cruising, I was advised to 'put your boat a couple of feet from the lower gates and you will hold her on a thread'. In the Bosley locks near our home moorings, there is practically no movement backwards or forwards as the water rises, but in the Trent & Mersey locks things are very different. As has already been noted, the ground paddles let in water half-way along the lock and it wells up and flows towards both the top and the bottom gates. Thus, if the boat is slightly forward of the centre, it may be washed with considerable force against the fender on the sill or against the top gate. The boat itself may stand the bump but nerves and china are likely to be shattered. A light boat may be held by a rope but a heavy one is practically impossible to stop. We therefore follow the advice and lie well back in a lock with a stern fender to keep us off the gates and a stern rope round a bottom gate beam for added security or nose the front of the lock with the engine in forward gear, watching the bow fender to see there is no obstruction. To avoid bumps, it is absolutely essential to get into position *before* the paddles are opened. Heaven preserve us from the enthusiastic helper who flings open the paddles before we are ready. He or she seldom realises that the ensuing catastrophe would not have occurred without such help.

The position in the front of the lock may have certain disadvantages. A leaky top gate can pour a lot of water into the boat if the front is open. On a narrowboat, the cratch is so arranged to deflect the fountain and keep the hold dry. Occasionally obstructions occur on the sill itself or on the gates and these will catch the rising bow. It is the duty of the one operating the paddles to see that this does not happen for the steerer is too far away. My daughter, Jennifer, once came upon a boat which had actually lifted a top gate from its seating. Luckily the boat was a heavy steel hull and was undamaged and it proved possible to reseat the gate and continue with the locking.

A completely different hazard in an ancient lock is a projecting course of brick or stone under which one side of the boat may catch. This actually happened to us both in Grindley Brook and in Hillmorton and prompt action was necessary to drop paddles and lower the water level. The operator on the lock must never be far away either physically or mentally!

Wide locks have their own problems. Once more there are few difficulties locking down provided the boat is kept well clear of the sill. There are usually bollards along the sides and a stern rope can be passed round one and held by the steerer. It must not be looped round in case it fails to run easily and knots in such a way as to prevent the stern from falling with the water. A heavy boat would snap the rope but a light one could be tilted so far forward that it would eventually take water and sink.

Rising through a wide lock it is advisable to secure the boat to a bollard. Many wide locks, particularly the deep ones at the eastern end of the Trent & Mersey, draw the boat forward. I usually pass the rope round the bollard and hand the end back to Pearl and she has little difficulty in holding and taking in the line as the boat rises. There may be a choice of bollards and we used to take the one at the back of the lock being careful not to catch the stern under a cross-piece of the gate. On Grand Union locks, the lock-keeper advised us to use a bollard farther forward and this has held us nicely. With no bollards, we tie the stern rope to a bottom gate beam. The order of raising the paddles is important in a wide lock when the boat is going up unaccompanied. The ground paddle and then the gate paddle should be raised on the same side as the boat as the water surges across the lock and rebounds off the wall to hold the boat in to its own side.

Perhaps the secret of all successful locking is to take time and go gently until the locks are thoroughly understood. A great deal of anxiety and some actual damage can be avoided by careful thought and a firm decision not to be rushed either by keen members of the crew or by over-enthusiastic helpers. The great flight at Audlem is innocuous once the boat has passed the poorly sited overflows. The mighty Hatton flight is much easier than it looks if proper care is taken. The Bingley five and the other Leeds & Liverpool staircases can bump a boat around unless it is properly held. It is all a question of care and patience and following the rules until with experience we can say with the confidence of the boatman we met on the Dudley canal, 'I'm familiar with these locks.'

CHAPTER TWELVE

THE CHESHIRE RING

In *Water Rallies* I described the long campaign to reopen the Cheshire Ring to navigation. The series of canals forming this ring include the Macclesfield from Marple to Kidsgrove and the Trent & Mersey from Kidsgrove to Preston Brook, near Warrington. These two canals have been navigable and well used throughout. The Bridgewater from Preston Brook to Castlefield, Manchester, has also been well maintained but has recently been breached at Dunham Town, near Altrincham. From Castlefield there is a short length of the still navigable Rochdale canal in the heart of Manchester and the Ashton and Lower Peak Forest canals through Fairfield and Dukinfield lead to the splendid flight of sixteen locks at Marple. These last two canals, still legally navigable, were last cruised in their entirety in 1961. The frost damage to the great aqueduct over the river Goyt at Marple in 1962 caused this to be de-watered and the Marple locks quickly deteriorated. Finally the leak in the Store Street aqueduct in Manchester, at the western end of the Ashton canal, sealed off the whole 15 mile length and its condition has steadily become worse.

I described how the Marple aqueduct was restored and repaired and how the National Rally of Boats at Marple and in Manchester in 1966 focused attention on the severed link so that public clamour would call for the complete repair and reopening. So popular was the rally of boats at Marple that annual local rallies were called for on the same beautiful site and were organised with great success from 1967 to 1970.

When the Peak Forest Canal Society was founded in 1964 we approached the British Waterways Board for permission to clear out the Marple locks and repair them with volunteer labour. This was not granted for the Board had little money to spare for these two lengths of canal and could not then consider restoration. However, towpath tidying lower down was allowed and this enabled the society to form a working party.

After a couple of years, with great local support, the society decided to take matters into its own hands and start clearing and repairing the Marple locks. It appeared to be the only possible way to stop vandalism which threatened to wreck the whole flight. A keen enthusiastic working party of volunteers collected each weekend and rapidly advanced in ability and experience. They were still without permission to work on the locks and in due course a letter came from the Board which pointed out that they were trespassing. Undeterred, they replied that they were the only trespassers who were improving the canal and they asked to be allowed to purchase and fit new lock beams. This was never formally agreed but in due course the work was carried out to everyone's satisfaction and repairs were made, lock by lock, until the whole length held water once more.

In 1968 the Board, now newly constituted, realised the firm intentions of the society and offered co-operation over paddle gear. This was accepted gratefully and the work continued.

The year 1968 was the year of the Transport Act when lengths of canal throughout the country were designated 'Cruiseways' to be kept and maintained for amenity purposes. The Macclesfield and Trent & Mersey canals were on the list, but the Lower Peak Forest and Ashton canals were not as their future was yet to be decided, very largely on the wishes of the local people along the line of the canals. This was reasonable and our rallies had been aimed at publicising the value of these canals to the local authorities. Unfortunately for us they were divided into two camps, those on the Lower Peak Forest wanting restoration of the whole length and those on the Ashton being more interested in abandonment. We could understand the Ashton canal authorities for the whole canal was in a bad way and looked a terrible mess. It was often referred to as a 'killer canal' for child drownings occurred in derelict regions where no one overlooked or cared. We knew, however, that abandonment was not the answer for derelict canals do not disappear. So much water was carried and delivered to the thirsty industries of Manchester that simple filling-in was not feasible without the laying of pipes and culverts. Further, we knew from actual culverting of other urban canals that the cost was enormous, greatly exceeding £100,000 per mile.

Under the Transport Act of 1962, the British Waterways Board had a duty to keep these canals in a condition at least as good as it had been in a six-month period of 1961 and in that period one boat at least had navigated the whole length. This they had not done and it was decided by a number of local authorities and canal users to take legal action against the Board. The attorney-general accepted a fiat to act on their behalf but the law is slow and before the case could come to court, the 1968 Transport Bill went to Parliament taking away the right of legal action to enforce maintenance. The Bill was a long one and took up so much time that the government applied the guillotine. In the House of Lords an amendment was passed allowing the legal action to take place but when the Bill returned to Commons the amendment was removed. Thus, in the autumn of 1968, we still had two canals usable only by small trailed boats. It was at this point that 'Operation Ashton' took place.

In 1967 a little publication appeared called *Navvies Notebook* and its editor, Graham Palmer, was the rare combination of an idealist and a practical man who saw the value of co-ordinating the work of volunteers throughout the country. This booklet recorded the volunteer work being carried out in different places and noted future programmes of work. It was a great encouragement to working parties and very soon groups from different areas were taking trips to help others. Furthermore, the number of working parties grew and each party became a focus for more volunteers.

In September 1968 it was decided to mobilise the workers and tackle the top mile of the Ashton canal locks from Clayton Junction, where the Stockport branch entered, to the top lock at Fairfield. A great operation was organised in military detail to take place on 21 and 22 September. The purpose was to clear as much rubbish as possible, to demonstrate to local authorities on the spot that their canal could easily become navigable once more and would be greatly improved with maintenance, and to take positive action with a problem which had lain so long unsolved.

The preparations were largely carried out by the Peak Forest Canal Society, advised by Graham Palmer, which sent in a working party beforehand. The British Waterways Board co-operated by lowering the water level and emptying the canal as far as possible. Plans were made for the feeding and sleeping of the volunteers and a headquarters marquee was set up on a piece of open ground. Dumpers and various items of heavy equipment were hired and the corporation departments of Manchester and Droylsden offered the use of their dumps for rubbish.

The Saturday morning of the start of the operation dawned cold and grey with rain driving in from the south-west. Pearl, who had volunteered for cookhouse work, Roger and I, ordinary labourers, put on our oldest clothes and set off to the site. We parked the car and were directed to lock 13 at Crabtree Lane.

The canal, largely emptied of water, as we saw it that day looked a foul and miserable rubbish dump. Our lock and all the others we saw contained tons of rubbish piled against the sill and lying deep in mud along the whole chamber. It had been this rubbish which made the canal so dangerous to children for much of it stood above the water forming slippery islands surrounded by deep channels and pools. All along the intervening pounds lay more rubbish of every description and part of the masonry of some of the locks which had been pushed in as well.

As we reached lock 13 a ladder arrived and we descended into the chamber. We had brought some rope and soon found an old dustbin and we set to work to haul out the rubbish piece by piece. For the next three hours we saw little but our own party of volunteers, now quite a large one, and the rubbish steadily built up at the lockside. All the while it poured with rain but we were soon so wet that we scarcely noticed it. We all agreed afterwards that these conditions, which so impressed the public, were not unpleasant for working. The onlookers remarked, however, that if volunteers could work in such conditions then restoration of the canal must be worth while.

It was when we stopped for lunch and walked back along the towing path to the headquarters marquee that we realised the size and success of the operation. Each society or working party had been given a particular job to do, a lock to clear or a stretch of intervening pound to tidy. Most of them had brought notices to say who they were—'The Wolverhampton Boat Club' 'The Midlands I.W.A.', 'The London and Home Counties I.W.A.' to name but a few. There were parties from Southampton and from Grimsby, though neither of these places has a canal of its own. There were groups from all over the country and the total number of volunteers was said to be around 600. Everyone was happy, cheerful and keen despite the conditions for they knew they were doing something worth while. All along the muddy towpath the dumpers were carrying away the rubbish.

In the marquee were piles of freshly cut sandwiches and gallons of hot tea and coffee—enough for all. Someone had provided a free barrel of beer, but the workers were keen to get back into the cut.

Back at lock 13 bonfires were being started. Though everything was sodden we soon had old rubber tyres in flames and all that was burnable was soon dry enough to be consumed. The rubbish that would not burn went into the dumpers and was carried to one of two huge dumps, there to be loaded into lorries and carted away. All kinds of things turned up in the mud. There was a brand-new motor-cycle, probably stolen, which the police

M

collected. A safe with its back blown off appeared and an un-exploded anti-aircraft shell. One of the policemen from the little borough of Droylsden suggested that if we found a body we should push it over the boundary into Manchester where they would have more people to deal with it! Fortunately the eventuality did not arise. In all, we were told afterwards, some 2,000 tons of rubbish was removed and this would not count all that was burnt.

The work was not restricted to rubbish removal. Where the locks were damaged the stones were carefully picked out and set on one side and on the Sunday, when the sun appeared, restoration was started on the damaged sections. One pair of lower lock gates which were broken beyond repair were removed by an 8 ton crane.

The attitude of the local people was most encouraging and one young man joined our party. He had come to see and he stayed to help. This happened all along the cut and we heard nothing but pleased comment on what we were doing. One man said he had wanted the canal abandoned but as none of the authorities ever did anything he was now on the side of the volunteers.

When work finished on the Saturday we went home and changed, then returned to the scout hut where the ladies had produced a huge and tasty stew followed by apple pie. Hundreds of people sat down to enjoy it and later most of them laid out their bedding on the floor. In the evening there was a sing-song at one of the 'locals'.

Except for one sharp thunderstorm, Sunday was fine. Throughout the day the work proceeded until four o'clock when it was time to clear up and collect the equipment. Lorries were still carting away the rubbish from the main dumps until the middle of the week. As we walked back along the towpath we were all struck by the transformation. The towpaths were still very muddy but the channel and locks were almost entirely clear of rubbish. Had all the gates been intact, the passage through the length would have been easy. Though there was a great deal of silt by the banks, the navigation channel would have been 4ft to 5ft deep all the way. Much more important, the dangerous rafts of rubbish in and near the locks had now been removed. Whatever else was to happen in future, we had made this length of the canal less dangerous for children. A week later, with the water levels back in all the stretches where the lock gates would hold, the canal was transformed.

Operation Ashton had been an ideal. It had shown that there were hundreds of people prepared to work voluntarily for something they believed in. It had shown that they could be mobilised to carry out a prodigious task and Graham Palmer of *Navvies Notebook* together with the talented Peak Forest Canal Society's committee has achieved wonders in the organisation of this huge operation. It demonstrated the latent power behind canal restoration, ready to be called upon for any worth-while job.

The cost? 'Operation Ashton' certainly cost money and over £1,000 was needed to pay for hire of lorries, dumpers and a host of other expenses. It was all subscribed voluntarily by canal lovers and when the last accounts were paid there was still money in hand. Thus the volunteers demonstrated to the authorities how best to deal with our less attractive and semi-derelict waterways. Never again could it be said that a task was too big to tackle.

After this great mass effort it was decided to concentrate on the local authorities and show them the true facts. The Peak Forest Canal Society working parties continued to clear and improve sections of the two canals and made another special effort on the Marple locks and side ponds in the autumn of 1969. At the end of that year and in early 1970 we were much heartened by a visit from Alderman Illtyd Harrington, Chairman of the Inland Waterways Amenity Advisory Council, and some of his colleagues, whose report advised complete restoration. We held meetings in the canal-side towns and a specially large one in Manchester. Though they were public meetings, we seldom seemed to see or hear anyone who really opposed restoration. Further, we felt that we were making little headway with the elected representatives who would have much to say in the final decision, so two special meetings were planned. One was a visit to Birmingham to see canal developments at Farmers Bridge, planned by Peter White, then on the architectural staff of Birmingham Corporation. Here a new housing scheme had made a feature of the waterway and British Waterways had co-operated in tidying up the canal. Special features were made of the attractive canal-side buildings and a brewery had built the *Longboat*, a delightful pub. We booked a boat to enable our visitors to cruise into the area and then issued invitations to aldermen, councillors, and council officers. The day trip was to be at our expense but we wondered how many would accept. In the event they responded wonderfully and we needed a second boat. Despite bitterly cold weather they were all most impressed.

The other meeting was a two-day conference in Manchester to which we invited the council representatives once more. Again it was well attended and we discussed the problems of urban canals. Our local authority colleagues clearly wanted the facts and in the end they were largely on our side. We published the talks and the discussions and circulated them to all canal-side authorities.

Another factor appeared at this time to help our cause. The Rochdale canal above Dale Street, abandoned in 1952, was to be turned into a shallow water channel in which children could play and paddle and which would form a small linear park. It was only $2\frac{1}{2}$ miles long but the scheme was to cost over half a million pounds—three times as much as restoring the whole 6 miles of the Ashton canal. This high cost carried weight with those who would have to find the money.

In the autumn of 1970 John Heap, Chairman of the North West branch of the IWA, called together a committee to plan a cruise up the Rochdale canal into Dale Street basin, right in the middle of Manchester. If 100 boats could be brought up this canal for a weekend, we should highlight once more the feasi-bility of through navigation. The immediate question was whether we could get one, let alone 100 boats through.

In *Water Rallies* I described how we brought narrowboat *Parrot* through the nine locks, the journey taking eleven days spread out over four weeks. In 1966, we had brought thirty-eight boats through the five lower locks at the time of the National Rally of boats at Marple. A few other trips had been made but the locks had deteriorated and repairs would be necessary. We approached the Rochdale Canal Company for their views and were received with great friendliness and courtesy. They would do all in their power to help us, their engineer, Mr Kay, would join our organising committee and we could hold our meetings in their offices. We should need to do much of the clearing and repairing work ourselves and it would be a co-operative effort. By this time the Peak Forest Canal Society working party had become a highly skilled body and they set to work to carry out a complete survey of the mile and a quarter length and its nine locks. Their report showed that very extensive repairs would be necessary, including the replanking many of the gates, replacing of some lock beams and repairs to paddle gear. It was not to be a simple case of pulling out rubbish but a task of restoration. Working each weekend from Christmas to Easter they considered that they could do it. Our plans went ahead.

The working party decided to start at the bottom lock and work upwards. The bottom lock gates were known to be shaky and it then appeared that this would be the most difficult lock to repair. All others could be emptied by draining the pound below, but this lock opened on to the longest level pound in the country and it would therefore be necessary to put in stop planks below the bottom gates and pump the water out. We were told that no stop planks had been inserted for at least a quarter of a century and that there might be considerable difficulty. However, they were quite essential for the repair to be able to take place.

Stop planks for a 14ft 6in width of canal are very large, heavy and awkward. When the first was inserted into the grooves it floated on the surface. The second pushed it down the grooves but floated higher out of the water and later ones floated higher still, so that it became almost impossible to force the whole lot on to the canal bed. The volunteers worked all evening until midnight, as many as fifteen standing on the top plank to add their weight. Miraculously no one fell in. It was not until the next day when the planks had become waterlogged that they began to fit and the 6in pump could keep up with the water which came in from both sides. The repair work then started in earnest and the gates were stripped of rotten timber and re-planked. There was a great mass of rubbish in the lock and this was loaded into boats lent by the Bridgewater Canal Company.

From that weekend the work proceeded steadily. It was hard and specialised with long hours of work in poor conditions but the volunteers pressed on. Some weekends they were joined by groups from the cruising clubs and other workers from much farther afield, but the brunt of the work fell on a small caucus of very experienced workers who were always available. In par-ticular Timothy Noakes, a student from the university, was out-standing both as organiser and craftsman. Peter Stockdale and Ian McCarthy were two more and several others were of similar calibre.

Things went reasonably well until they came to the penulti-mate lock, the lock beneath the eighteen storey Rodwell Tower, so difficult for *Parrot* seven years earlier. The lower gates were found to be in a state of collapse and one, in fact, started to break up as they swung it. They commented afterwards that had they known the state of this lock at the start, there would have been no rally. They were joined by Dr Paul Spriggs, a metal-lurgist and metal worker, and in one weekend of almost con-tinuous work they pulled and strapped the gate together with metal rods. At this stage they were camping under the skyscraper.

The basins themselves needed clearing out and British Water-ways permitted the preparation and use of the Ashton canal basins as well. By now we had over one hundred entries and we began to wonder where the boats would be moored if they ever reached the basin. The Rochdale Canal Company had offered us the use of a fine early nineteenth-century warehouse for the rally itself and this we set to work to clean out and white-wash. By Easter all the last-minute jobs were being completed and many boats had started from their moorings for the rally cruise. The indefatigable Eric Wilson had fixed electric lights over the Rodwell Tower lock and we reckoned we were ready—*if* we could keep up the water supplies.

It had been decided that the first boats would be allowed through the bottom lock at Castlefield on the Thursday after Easter about midday. Pearl and I were to join the little party at Castlefield to issue plaques and to help with the locking through. It was about eleven o'clock when we arrived on the scene on a rather damp and rainy day and were promptly invited into what used to be the lockhouse by the householders Mr and Mrs Roberts who placed a room at our disposal. Three boats had been allowed up the evening before and there were a number more waiting. The message came down that water was already short and no more boats were to come up until supplies were more certain. The Rochdale canal above Dale Street was aban-doned to navigation in 1952 and a 2-mile stretch was being 'landscaped' into a series of shallow pools. While the work was proceeding, the water was being carried over the section in a pipe. The canal company was sending water down from the reservoirs and it was seen to enter the pipe but nothing came out at the other end. After careful inspection and frantic tele-phoning, this was put right. Just before three, we sent up a batch of four boats but three more hours elapsed before the next lockful could start. Three further lockfuls were sent up before we finally finished for the night.

Next morning, thanks largely to British Waterways who put as much water as they could through the Ashton canal, we had boats through before nine and carried on at half-hour intervals throughout the day. By now boats were moored as far as we could see along the Bridgewater waiting their turn and we checked them for size and packed them in as tightly as we could. Once a group of boats was put in to the bottom lock they stayed together throughout the flight and every lock was worked for them.

Our lock required great care for it was the only one without ground paddles and the large gate paddles were particularly fierce. To avoid being swamped, boats were instructed to take a bow rope round a bollard and keep it taut throughout. To make matters more difficult the locksides were cut out of rock which was slightly undercut and we had to ensure that no gunwales were jammed beneath the projections. All was well when boats were held in firmly but one or two swung out and received a deluge over the bows before we could right them. However, all boats passed through without accident.

At eleven o'clock the TV cameras arrived and went aboard *Duchess*, the chairman's boat. The newspaper reporters and cameras came through as well and gave us good publicity. At half-past four I launched my little punt *Rosebud* built specially during the winter and went through myself; 8ft long and 2ft beam, she was the smallest boat in the rally and had the honour in being numbered 'one'. Throughout the afternoon and evening, we worked on and put the last boats through just after eight o'clock. On Saturday morning we started early and continued to lock boats through until all were in the Rochdale canal. At twelve noon there were no more waiting below and we closed the lock and padlocked the gates. We had worked 103 boats through, filling and emptying the lock 26 times and used about 2 million gallons of water including that lost through the leaks in the gates.

We were not to see the actual site until the Saturday afternoon but the work of fitting every boat into the confined spaces of the Rochdale and Ashton basins was as complicated a jig-saw puzzle as had ever been conceived. It was successfully achieved by Ken Goodwin, the harbour-master, using every inch of space. People said afterwards that they had never had a worse rally mooring but had never heard so little grumbling. All were so thrilled to have made the journey and reached the site.

We had planned a special VIP cruise for the Saturday afternoon. The Lord Mayor of Manchester and several neighbouring mayors and chairmen of councils had agreed to join us at the rally site and we planned to take them aboard *Clevanda* and *Duchess*, the boats belonging to John Humphries, the IWA chairman, and John Heap, the rally chairman. They were to be given tea and taken on a short cruise through two locks and beneath the huge Rodwell Tower into the centre of the rally. The guests were to come aboard at four o'clock and crowds of people awaited their arrival both on the towpath and on the rally site. All went well according to plan and the Lord Mayor spoke from the microphone and told us how he would like to see the whole of the Cheshire ring of canals reopened. In the evening, we enjoyed a Victorian melodrama in the beautifully spruced up warehouse.

Sunday dawned fine and bright and we were faced with the task of getting the boats out again. It had taken two days to bring them up the canal and during that time the water levels had seemed dangerously low at times. The first boats entered the top lock a few minutes before nine o'clock. To our delight everything went smoothly and by half-past five the last of the hundred passed through Castlefield lock and out into the Bridgewater canal. The Rochdale canal was fully navigable once more and the credit must be divided between the highly skilled Peak Forest working party and the enterprising canal company.

At about this time we learned that the canal-side authorities had at last agreed amongst themselves to press for the complete restoration of the Ashton and Peak Forest canals and were now about to apportion the costs between themselves and the British Waterways Board. It seemed now that our long fight of more than ten years for the future of the Cheshire Ring was won. We were able to picture setting out from our home moorings at Higher Poynton to Marple where we should drop down through sixteen deep locks and cross Outram's fine aqueduct. We should then cruise through the woodlands and continue to Dukinfield junction which was dominated by the graceful Ashton canal warehouse. At Fairfield we should descend eighteen more locks and pause in the marina planned by the Rochdale Canal Company in Dale Street. Finally, we should cruise through the heart of Manchester on the Rochdale canal and join the Bridgewater at Castlefield where we had helped to lock through so many boats. Once there, we could make for the Leeds & Liverpool or return home via the Trent & Mersey and the Macclesfield canal. For so long, this had been only a dream, but at last it was to become a reality.

In August of 1971, the Cheshire Ring was to receive yet another devastating blow. The great embankment of the Bridgewater adjoining the Bollin aqueduct burst its banks and collapsed over a considerable length. Millions of gallons of water poured into the fields and the river below before stop planks could be inserted to seal off the longest stretch of level water in the country. Boats were left on the mud and others away on holiday were cut off from their bases. We were at Northampton at the National Boat Rally at the time, moored near members of the Sale Cruising Club who told us despondently that they had no easy way home. They could go the long way via the Leeds & Liverpool or satisfy the rigorous conditions of the Manchester Ship Canal, but their direct route was cut. Furthermore the Cheshire Ring was broken and the physical break was a major one.

The politics were highly complicated. The Bridgewater belonged to the Manchester Ship Canal Company who could not be expected to welcome a large expenditure on a subsidiary which they considered to be without profit to them. They had just missed being nationalised and this would have placed the Bridgewater under the British Waterways Board who were deeply concerned at the effect of the breach on their adjoining canals. It was a great relief when the Ship Canal Company decided on complete restoration.

So the long story of the troubles of the Cheshire Ring has been concluded and we are sure that we shall eventually fulfil our dream to cruise the whole ring in *Rose of Sharon*.

REFERENCES

Bliss, W. *The Heart of England by Waterway* (Witherby, 1933)

Court, W. H. B. *The Rise of the Midlands Industries 1600–1838* (OUP, 1938)

Ekwall, E. *The Concise Oxford Dictionary of English Place Names*, 4th ed (OUP, 1966)

Forester, C. S. *Hornblower and the Atropus* (Michael Joseph, 1954)

Fournel, H. & Dyévre, I. *Memoir of the Underground Canals and Coal Mines of Worsley, near Manchester* (Paris, 1842)

Haden, H. J. *Notes on the Stourbridge Glass Trade* (Library & Arts Committee, Brierley Hill, 1949)

Hadfield, C. *The Canals of the East Midlands*, 2nd ed (David & Charles, 1970)

Hadfield, C. *The Canals of the West Midlands*, 2nd ed (David & Charles, 1969)

Hadfield, C. *The Canals of South and South East England* (David & Charles, 1969)

Hadfield, C. & Biddle, G. *The Canals of North West England* (David & Charles, 1970)

Hadfield, C. & Streat, M. *Holiday Cruising on Inland Waterways* (David & Charles, 1968; Pan Books, 1972)

Heath, J. E. *A History of Long Eaton* (Long Eaton Council, 1967)

Household, H. *The Thames & Severn Canal* (David & Charles, 1969)

Jackman, W. T. *The Development of Transportation in Modern England*, 2nd ed (Cass, 1962)

Owen, D. E. *Water Highways* (Dent, 1967)

Owen, D. E. *Water Rallies* (Dent, 1969)

Phillips, J. *A General History of Inland Navigation*, 5th ed (1805, reprinted David & Charles, 1970)

Priestley, J. *Historical Account of the Navigable Rivers, Canals and Railways of Great Britain* (1831, reprinted David & Charles, 1969)

Rolt, L. T. C. *Narrow Boat* (Eyre & Spottiswoode, 1944)

Rolt, L. T. C. *The Inland Waterways of England* (Allen & Unwin, 1950)

Salis, H. R. de. *Bradshaws Canals and Navigable Rivers* (1904, reprinted David & Charles, 1969)

Seymour, J. *Voyage into England* (David & Charles, 1966)

Slough Canal Study Group. *Slough Canal—the Future?* (1969)

Stevenson, P. *The Nutbrook Canal* (David & Charles, 1970)

Thurston, T. *The Flower of Gloster* (1911, reprinted David & Charles 1969)

Wilkinson, T. *Hold on a Minute* (Allen & Unwin, 1965)

Willan, T. S. *River Navigation in England 1600–1750* (1936, reprinted Cass, 1964)

Willan, T. S. *The Navigation of the River Weaver in the Eighteenth Century* (Chetham Society, 1951)

INDEX